Wissenschaftliche Reihe Fahrzeugtechnik Universität Stuttgart

Reihe herausgegeben von

Michael Bargende, Stuttgart, Deutschland

Hans-Christian Reuss, Stuttgart, Deutschland

Jochen Wiedemann, Stuttgart, Deutschland

Das Institut für Fahrzeugtechnik Stuttgart (IFS) an der Universität Stuttgart erforscht, entwickelt, appliziert und erprobt, in enger Zusammenarbeit mit der Industrie, Elemente bzw. Technologien aus dem Bereich moderner Fahrzeugkonzepte. Das Institut gliedert sich in die drei Bereiche Kraftfahrwesen, Fahrzeugantriebe und Kraftfahrzeug-Mechatronik. Aufgabe dieser Bereiche ist die Ausarbeitung des Themengebietes im Prüfstandsbetrieb, in Theorie und Simulation. Schwerpunkte des Kraftfahrwesens sind hierbei die Aerodynamik, Akustik (NVH), Fahrdynamik und Fahrermodellierung, Leichtbau, Sicherheit, Kraftübertragung sowie Energie und Thermomanagement – auch in Verbindung mit hybriden und batterieelektrischen Fahrzeugkonzepten. Der Bereich Fahrzeugantriebe widmet sich den Themen Brennverfahrensentwicklung einschließlich Regelungs- und Steuerungskonzeptionen bei zugleich minimierten Emissionen, komplexe Abgasnachbehandlung, Aufladesysteme und -strategien, Hybridsysteme und Betriebsstrategien sowie mechanisch-akustischen Fragestellungen. Themen der Kraftfahrzeug-Mechatronik sind die Antriebsstrangregelung/Hybride, Elektromobilität, Bordnetz und Energiemanagement, Funktions- und Softwareentwicklung sowie Test und Diagnose. Die Erfüllung dieser Aufgaben wird prüfstandsseitig neben vielem anderen unterstützt durch 19 Motorenprüfstände, zwei Rollenprüfstände, einen 1:1-Fahrsimulator, einen Antriebsstrangprüfstand, einen Thermowindkanal sowie einen 1:1-Aeroakustikwindkanal. Die wissenschaftliche Reihe „Fahrzeugtechnik Universität Stuttgart" präsentiert über die am Institut entstandenen Promotionen die hervorragenden Arbeitsergebnisse der Forschungstätigkeiten am IFS.

Reihe herausgegeben von

Prof. Dr.-Ing. Michael Bargende
Lehrstuhl Fahrzeugantriebe
Institut für Fahrzeugtechnik Stuttgart
Universität Stuttgart
Stuttgart, Deutschland

Prof. Dr.-Ing. Jochen Wiedemann
Lehrstuhl Kraftfahrwesen
Institut für Fahrzeugtechnik Stuttgart
Universität Stuttgart
Stuttgart, Deutschland

Prof. Dr.-Ing. Hans-Christian Reuss
Lehrstuhl Kraftfahrzeugmechatronik
Institut für Fahrzeugtechnik Stuttgart
Universität Stuttgart
Stuttgart, Deutschland

Sebastian K. Crönert

A Complete Methodology for the Predictive Simulation of Novel, Single- and Multi-Component Fuel Combustion

Springer Vieweg

Sebastian K. Crönert
IFS, Chair in Automotive Powertrains
University of Stuttgart
Stuttgart, Germany

Zugl.: Dissertation Universität Stuttgart, 2023
D93

ISSN 2567-0042 ISSN 2567-0352 (electronic)
Wissenschaftliche Reihe Fahrzeugtechnik Universität Stuttgart
ISBN 978-3-658-43074-0 ISBN 978-3-658-43075-7 (eBook)
https://doi.org/10.1007/978-3-658-43075-7

This Springer Vieweg imprint is published by the registered company Springer Fachmedien Wiesbaden GmbH, part of Springer Nature.
The registered company address is: Abraham-Lincoln-Str. 46, 65189 Wiesbaden, Germany

Paper in this product is recyclable.

„Ne discere cessa!“
Marcus Porcius Cato

Preface

This work has been created during my employment as a research assistant at the Institute for Automotive Engineering Stuttgart at the University of Stuttgart under the direction of Prof. Dr.-Ing. M. Bargende. I owe him my special thanks for the supervision during the creation of this work. His remarks and ideas shaped the fundamental idea and helped to develop it to the current state. Great thanks also goes to Prof. Dr. techn. H. Eichlseder, and Prof. Dr.-Ing. A. Casal Kulzer who both volunteered to co-examine my work by joining the doctorate committee.

Additionally, i want to express my great gratitude for the help of Dr. Michael Grill. He helped me guide this thesis into the right direction, constantly supporting my ideas and visions. Special thanks also goes to my dear friend and colleague Mr. Sebastian Welscher for his help. I did greatly profit from the professional collaboration with him, especially in regards to the coding aspects of this work and the application of cloud computing. I would also like to thank all my other colleagues at IFS and FKFS.

The support of the FVV provided by funding the "Fuel Composition for CO_2 Reduction" project is greatly appreciated. A big thanks also goes to my project colleagues Mr. Burkhard, Mr. Hesse, Mr. Jacobs, and Mr. vom Lehn from RWTH Aachen University and Mr. Franken from BTU Cottbus-Senftenberg.

Lastly, I want to highlight the great support I was granted by my family and my partner. Without the help of my parents, especially in the early phases of my university experience, I would not have been able to achieve the things I now have. The relationship with my brother always inspired me and the healthy competition has elevated both our ambitions.

Stuttgart — Sebastian Karl Crönert

Table of Contents

List of Figures

List of Tables

Acronyms

$AN_{3.70oxy}$	TRFE5/AN mixture with 3.7 % (m/m) Oxygen content
$AN_{5.00oxy}$	TRFE10/AN mixture with 5.0 % (m/m) Oxygen content
$CPN_{3.70oxy}$	TRFE5/CPN mixture with 3.7 % (m/m) Oxygen content
$CPN_{5.00oxy}$	TRFE10/CPN mixture with 5.0 % (m/m) Oxygen content
$DMC_{65}MF_{35V}$	65/35 % V/V mixture of DMC with Methyl Formate
$E10AN_{50V}$	TRFE10 in a 50 % (V/V) mixture with Anisole
$E10AN_{50m}$	TRFE10 in a 50 % (m/m) mixture with Anisole
$E10CPN_{50V}$	TRFE10 in a 50 % (V/V) mixture with Cyclopentanone
$E10CPN_{50m}$	TRFE10 in a 50 % (m/m) mixture with Cyclopentanone
0D	Zero-Dimensional
1D	One-Dimensional
3D	Three-Dimensional
AFR	Air-to-Fuel Ratio
AI	Auto-Ignition
AN	Anisole C_7H_8O
BC	Boundary Condition
BMBF	Federal Ministry of Education and Research
CA	Crank Angle
CCV	Cycle-to-Cycle Variations

COV	Coefficient of Variation
CPN	Cyclopentanone C_5H_8O
DES	Differential Equation System
DMC	Dimethyl Carbonate
E10	Gasoline with 10 % (V/V) Ethanol Content
E10 AN	TRFE10 with AN (5.0 % m/m Oxygen content)
E10 CPN	TRFE10 with CPN (5.0 % m/m Oxygen content)
E5	Gasoline with 5 % (V/V) Ethanol Content
E5 AN	TRFE5 with AN (3.7 % m/m Oxygen content)
E5 CPN	TRFE5 with CPN (3.7 % m/m Oxygen content)
EGR	Exhaust Gas Recirculation
EOC	End of Combustion
ER	Equivalence Ratio
FKFS	Research Institute for Automotive Engineering and Powertrain Systems Stuttgart
FL	Flammability Limit
FVV	Forschungsvereinigung Verbrennungskraftmaschinen e. V.
gHCCI	gasoline Homogeneous Charge Compression Ignition
GT	Gamma Technologies
H/C	Hydrogen/Carbon
HGD	Chair of High Pressure Gas Dynamics

HRR	Heat Release Rate
HT	High Temperature
ICE	Internal Combustion Engine
IDT	Ignition Delay Time
IFS	Institute for Automotive Engineering Stuttgart
IMEP	Indicated Mean Efficient Pressure
ITVTC	Institute for Technical Combustion
IVC	Inlet Valve Closes
IVK	Institute for Internal Combustion Engines and Automotive Engineering
IVO	Inlet Valve Opens
KLSA	Knock Limitation Spark Advance
KPP	Knock Peak-To-Peak
LBV	Laminar Burning Velocity
LFL	Lower Flammability Limit
LHV	Lower Heating Value
LTC	Low Temperature Chemistry
MF	Methyl Formate
MFB	Mass Fraction Burnt
MON	Motor Octane Number
NaMoSyn	Nachhaltige Mobilität mit Synthetischen Kraftstoffen
NTC	Negative Temperature Coefficient

PRF	Primary Reference Fuel
PTA	Pressure Trace Analysis
QD	Quasi-Dimensional
RCM	Rapid Compression Machine
RON	Research Octane Number
RWTH	North Rhine-Westphalia Technical University of Aachen
SI	Spark Ignition
ST	Shock Tube
STP	Standard Temperature and Pressure
TDC	Top Dead Center
TME	Chair of Thermodynamics of Mobile Energy Conversion Systems
TPA	Three Pressure Analysis
TRF	Toluene (Primary) Reference Fuel
TRFE10	see TRF and E10
TRFE5	see TRF and E5
UFL	Upper Flammability Limit
WR	Water Rate

Glossary

Greek Letters

α	Calibration Constant LBV Correlation	-
β	Calibration Constant LBV Correlation	-
β_π	Beta-parameter for the calculation of Π_{AI}	-
δ_L	Laminar Flame Thickness	m
δ_t	Thermal Boundary Layer Thickness	m
η	Dynamic Viscosity	$Pa \cdot s$
κ	Reaction Rate Constant	$\frac{M^i}{s}$
λ	Stoichiometric Air-to-Fuel Ratio	-
λ_{th}	Thermal Conductivity Coefficient	$\frac{W}{m \cdot K}$
$\mathcal{L}_b$	Markstein Length	m
ν	Kinematic Viscosity	$\frac{m^2}{s}$
ϕ	Wilke Mixing Equation Constant	-
ϕ_{50}	Location of MFB50	°CA
ϕ_q	Heat Transfer	$\frac{W}{m^2}$
Π	Pi Criterion	-
ρ	Specific Density	$\frac{kg}{m^3}$
τ	Ignition Delay Time	s

Roman Letters

A	Arrhenius Frequency Factor	-
a	Correlation Coefficient NASA Polynomials	-
α_{δ_L}	Prefactor Laminar Flame Thickness Model	-
AFR	Air-to-Fuel Ratio	$\frac{kg}{kg}$
a_s	Speed of Sound in the Burnt Gas	$\frac{m}{s}$
B_i	Calibration Constant LBV Correlation	bar
B	Arrhenius Exponent	$\frac{MJ}{kg}$
b	Temperature Exponent, NASA Polynomial Coefficient	-
B_E	Engine Bore	m
b_{H2O}	Calibration Constant LBV Correlation	-
C	Temperature Increase Model Parameter	$\frac{1}{K^{4-i}}$
c	Molar Concentration	$\frac{mol}{m^3}$
c_{H2O}	Calibration Constant LBV Correlation	-
c_p	Specific Heat Capacity for Constant Pressure	$\frac{J}{kg \cdot K}$
CR	Compression Ratio	$\frac{V}{V}$
D	Diffusion Coefficient	$\frac{m^3}{s}$
Da	Damköhler-Number	-
D_F	Molecular Diffusivity of the Fuel	$\frac{cm^2}{s}$
E_i	Calibration Constant LBV Correlation	K
E_a	Activation Energy	J
F	Calibration Constant LBV Correlation	$\frac{cm}{s}$
f	Frequency	$\frac{1}{s}$

FL	Flammability Limit	$\frac{V}{V}$
G	Calibration Constant LBV Correlation	K
H	Overall Enthalpy	J
h	Species Enthalpy	J
$h2o_a$	Water Coefficient LBV Correlation Hydrogen	-
$h2o_b$	Water Coefficient LBV Correlation Hydrogen	-
HRR	Heat Release Rate	$\frac{J}{s}$
i	Iteration Integer	-
J	Conservation Value Flow	-
j	Diffusion Flow	$\frac{kg}{m^3 \cdot s}$
k	Linear Constant Factor	-
l	Length	m
Le	Lewis Number	-
LFL	Lower Flammability Limit	$\frac{V}{V}$
LHV	Lower Heating Value	$\frac{MJ}{kg}$
l_T	Taylor Microscale	m
M	Molar Mass	$\frac{kg}{mol}$
m	Mass	kg
Ma	Markstein Number	-
$mult$	Multiplier for LBV Correlation H2	-
n	Mole Count	mol
p	Pressure	bar
p_1	Polynomial Coefficient LBV Correlation H2	-
p_2	Polynomial Coefficient LBV Correlation H2	-
p_3	Polynomial Coefficient LBV Correlation H2	-

p_4	Polynomial Coefficient LBV Correlation H2	-
p_5	Polynomial Coefficient LBV Correlation H2	-
$p_{\mathrm{egr,add}}$	EGR Coefficient LBV Correlation H2	-
$p_{\mathrm{egr,log}}$	EGR Coefficient LBV Correlation H2	-
P	Lambda Polynomial Ignition Delay Correlation	-
Q	Heat	J
Q_{s}	Source Term	-
Q	EGR Polynomial Ignition Delay Correlation	-
R	Molar Gas Constant	$\frac{\mathrm{J}}{\mathrm{K \cdot mol}}$
r	Radius	m
RR	Reaction Rate	$\frac{\mathrm{mol}}{\mathrm{L \cdot s}}$
r_{s}	Species Source Term	-
R	Lambda Polynomial Ignition Delay Correlation	-
S	Entropy	$\frac{\mathrm{J}}{\mathrm{K}}$
s	Velocity	$\frac{\mathrm{m}}{\mathrm{s}}$
$S_1(Z^*)$	First Spline LBV Correlation	-
$S_2(Z^*)$	Second Spline LBV Correlation	-
$S_3(Z^*)$	Third Spline LBV Correlation	-
$S_4(Z^*)$	Forth Spline LBV Correlation	-
s_E	Engine Stroke	m
S	EGR Polynomial Ignition Delay Correlation	-
T	Temperature	K
t	Time	s
T_{a}	Temperature Coefficient LBV Correlation H2	-
T_{ex}	Temperature Coefficient LBV Correlation H2	-

$\mathcal{X}_{\text{Taylor}}$	Taylor Factor	-
U	Pressure Polynomial Ignition Delay Correlation	-
u'	Turbulent Fluctuation Velocity	$\frac{\text{m}}{\text{s}}$
V	Volume	m^3
v	Velocity	$\frac{\text{m}}{\text{s}}$
V_h	Cylinder Displacement	m^3
W	Work	J
w	Mass Fraction	$\frac{\text{kg}}{\text{kg}}$
W_c	Conservation Value Density	$\frac{\text{kg}}{\text{m}^3}$
x	Molar Fraction	$\frac{\text{mol}}{\text{mol}}$
Z	Compressibility Factor	-
z	Distance	m
Z^*	Mixture Fraction	$\frac{\text{kg}}{\text{kg}}$

Chemical Species

(A)	Concentration of Species A
C_5H_8O	Cyclopentanone
C_7H_{16}	n-Heptane
C_7H_8O	Anisole
C_7H_8O	Toluene
C_8H_{18}	iso-Octane
CH_3^+	Methyl Cation
CH_4	Methane
CO_2	Carbon Dioxyde

H_2	Hydrogen Molecule
H_2O	Dihydrogen Oxyde
H_2O_2	Hydrogen Peroxide
HO_2	Hydroperoxyl
N_2	Nitrogen Molecule
NO	Nitric Oxyde
NO_2	Nitrogen Dioxyde
NO_x	Nitrogen Oxydes (usually NO and NO_2)
O_2	Oxygen Molecule
$^{\bullet}OH$	Hydroxyl Radical

Subscripts

0	inner layer
A	Arrhenius pre-factor
ad	adiabate
AI	auto-ignition
air	air
B	Arrhenius exponent
b	burnt
bb	blow-by
bl	boundary layer
C	Temperature Increase Model Parameter
c	critical
comb	combusted, related to combustion products

cyl	cylinder
E	entraining
EGR	to EGR
eng	engine
etha	ethanol
ex	exhaust
F	flame, Fuel
f	forward, flame
fit	corresponding to a correlation fit
fuel	fuel
H2O	water
hep	n-heptane
high	high/higher
in	intake
incr	increase
int	integral
j	iteration integer
KB	knock boundary
knock	knock, knocking
kpp	knock peak-to-peak
L	laminar
lam	lambda
liq	liquid
low	low/lower
end	end

max	maximum
start	start
min	minimum
mix	mixture
oct	iso-octane
r	reverse, reactand
st	stoichiometric
T	turbulent
t	tip
therm	thermal
tol	toluene
u	upper
ub	unburnt
W	(Cylinder) Walls
z	compressibility

	Superscripts
0	unstretched
c	calibration constant LBV correlation
m	calibration constant LBV correlation
n	calibration constant LBV correlation
n_a	calibration constant LBV correlation
n_{EGR}	calibration constant LBV correlation
r	calibration constant LBV correlation

Abstract

With the Paris agreement and its increasingly constricting emission goals, greater effort must be put into Internal Combustion Engine (ICE) design to reach them. Among other measures, the implementation of synthetic fuels gains in attractiveness. Due to their low local CO_2-emissions, especially fuels with low carbon content are looked at. To reduce costs for the adaption of existing engines and the design of new ones, combustion simulation can be key. It can help to gather information on an engines behavior under changes of the fuel (composition) and/or boundary conditions. An application of a novel fuel that has a high lean burn limits will consequently require a redesign of the whole charging system to secure the provision of the needed air. To be able to accurately predict the engine behavior for a fuel change, an suitable methodology has to be found to integrate novel fuels into existing or adapted simulation models.

To achieve this, this work is based on existing ones that have shown that for certain fuels it is possible to predict their combustive behavior solely on their physiochemical properties. As the measurement for laminar flame speeds and flame thicknesses is bound to low temperatures and pressures, engine relevant data was acquired by the means of reaction kinetic simulations. The same principle applies for the acquisition of ignition delay times. With the integration of proper distributed computing and cloud computing possibilities, the calculation process of those reaction kinetic simulation could be scaled by at least an order of magnitude. While for the laminar burning velocities and the laminar flame thicknesses, existing correlations could be used, the one for the ignition delay times had to be adjusted to fit to the data of fuel mixtures with Gasoline. Additionally, the proposal for a laminar burning velocity correlation for hydrogen is provided.

For the automated process, the fuel properties that depend on reaction kinetics are calculated first, all other intrinsic values (specific heat capacities, liquid densities, thermal conductivities, and critical points) are gathered by applying mixing rules to the neat components attributes. The automated correlation fits for laminar flame speeds, laminar flame thicknesses and auto ignition data allow for a quick adjustment to new fuels and fuel mixtures without the need of any manual work or surveillance.

For the validation, six different engines have been used. They vary in geometry, use case and operating points. Additionally, twelve fuels or fuel mixtures have been used to evaluate the performance of the fully automated process: Gasoline, Cyclopentanone with varying mixing rates into Gasoline, Anisole with varying mixing rates in Gasoline, Methanol, a Dimethyl Carbonate and Methyl Formate mixture, and Hydrogen. Two combustion simulation models (*Damköhler* and *Peters*) and two knock onset models (*Fandakov* and *Hess*) have been used. For all engines and fuels, the prediction results for the burn rates met the test bench results with sufficient precision. This is valid for both used combustion models. With the knock occurrence models, it has showed, that while the *Fandakov* model is roughly able to predict the fuel related changes, the *Hess* model shows better performance once it is recalibrated for the new fuel. Overall, the presented methodology can be used for predictive engine simulations of all fuels for that accurate reaction kinetic mechanisms exist.

Kurzfassung

Durch das Pariser Abkommen und die damit verbundenen, immer schärferen Emissionsziele muss mehr Aufwand in die Entwicklung von Verbrennungsmotoren gesteckt werden, damit ebendiese Ziele auch erreicht werden können. Die reine Effizienzsteigerung der Verbrennung selbst ist dabei aber schon sehr weit fortgeschritten - weitere Verbesserungen sind hier nur mit sehr hohen Kosten realisierbar. Zunehmend findet aber auch die Verwendung von Synthetikkraftstoffen immer weiter Zuspruch. Durch ihre sowohl global, aber eben auch lokal sehr niedrigen CO_2-Emissionen sind vor allem Kraftstoffe mit sehr geringem Kohlenstoffanteil im Gespräch. Um die Kosten für die Adaption von bereits existierenden Motoren an diese Kraftstoffe oder für die vollständige Neuentwicklung von Verbrennungskraftmaschinen möglichst gering zu halten, ist die Motorensimulation entscheidend. Sie kann dazu verwendet werden, das Motorverhalten bei einem Kraftstoffwech-sel mit gleichzeitig geänderten Betriebsbedingungen vorherzusagen. Für neuartige Kraftstoffe (zum Beispiel Wasserstoff), die für ihre Verbrennung sehr große Sauerstoffmengen benötigen, muss unter Umständen das gesamte Aufladesystem angepasst werden, um die Luftversorgung sicherzustellen. Um die Verbrennung von neuartigen Kraftstoffen genau vorhersagen zu können, muss basierend auf bereits vorhandenen Simulationsmodellen ein neuer Prozess zur Integration von weiteren neuartigen Kraftstoffen geschaffen werden. Dabei liegt das Hauptaugenmerk auf Kraftstoffmischungen, die bei einem Drop-In in die bestehende Fahrzeugflotte relevant werden könnten. Dennoch soll eine ebenso einfache und problemlose Verwendung für Reinkraftstoffe gewährleistet werden.

Um dies zu erreichen, baut diese Arbeit auf bereits existierenden Ergebnissen für methanbasierte Kraftstoffe auf. Diese zeigen, dass es für bestimmte Kraftstoffverbindungen möglich ist, deren Verbrennungsverhalten rein unter Hinzunahme ihrer physikalischen und chemischen Eigenschaften vorherzusagen. Die dabei für die vollständige Vorhersage nötigen Kraftstoffdaten lassen sich sehr grob in zwei Gruppen gliedern.

Die erste Gruppe enthält allgemeingültige Kraftstoffeigenschaften, die entweder durch eine einzige Randbedingung (zum Beispiel: Temperaturabhängigkeit der Entropie) beeinflusst werden oder sogar vollkommen unabhängig von den Umgebungsbedingungen sind (Beispiel: Unterer Heizwert). Sie können für Kraftstoffmischungen meist sehr einfach über passende Mischungsregeln bestimmt und dann als einzelne Variable oder in trivialen Korrelationen gespeichert werden. Einzige Ausnahme sind hierbei die NASA-Polynome. Sie stellen zwar ebenfalls nur eine einfache Abhängigkeit von der Temperatur dar, durch die gemeinsamen Koeffizienten der drei Gleichungen ist der Abstimmungsprozess aber etwas aufwändiger. Zur ersten Gruppe gehören folgende Kraftstoffeigenschaften: Die (flüssige) Dichte bei Standardbedingungen ρ, die dynamische Viskosität ν, die Wärmeleitfähigkeit λ_{th}, der kritische Punkt (T_{c}, p_{c}), der spezifische untere Heizwert LHV, die Enthalpie H, die spezifische Wärmekapazität c_{p} und die Entropie S.

Die zweite Gruppe enthält Kennwerte, die von sehr vielen motorischen Randbedingungen beeinflusst werden. Diese Effekte von äußeren Wirkungsgrößen müssen dann für den späteren Einsatz in der Simulation ebenfalls abgedeckt werden. Im einfachsten Fall kann dies durch das Hinterlegen von höherdimensionalen (fünf oder mehr Dimensionen - je nach Anwendungsfall) Kennfeldern geschehen. Die Verwendung von tabellierten Daten bringt jedoch die großen Nachteile des hohen Speicherplatzbedarfs und der notwendigen Inter- und Extrapolation mit sich. Gerade Letztere kann eine enorme Verschlechterung der erreichbaren Genauigkeit und Rechengeschwindigkeit mit sich ziehen. Um dies zu umgehen, werden spezialisierte Korrelationen erstellt, deren Kalibrierwerte automatisch an die zuvor durch Reaktionskinetikrechnungen ermittelten Sollwerte angepasst werden. So müssen später zur Laufzeit der eigentlichen Simulationsrechnung nur noch algebraische Gleichungen gelöst werden, Inter- und Extrapolationen sind bereits in ihnen enthalten. So wird ein sehr schneller Zugriff auf die Daten bei gleichzeitig hoher Genauigkeit erreicht. Durch die gezielte Auslegung der Korrelationen kann außerdem gewährleistet werden, dass die Randbedingungsbereiche, die für den motorischen Betrieb eine besondere Bedeutung haben, später die höchste Genauigkeit aufweisen. Detailliertere Korrelationen fordern: Die laminaren Flammengeschwindigkeiten s_{L}, die laminaren Flammendicken δ_{L} und die Zündverzugszeiten τ.

Da für die laminaren Flammengeschwindigkeiten und die laminaren Flammendicken prinzipbedingt nur Messdaten für vergleichsweise geringe Drücke und Temperaturen zur Verfügung stehen, werden die benötigten Daten bei motorrelevanten Bedingungen mithilfe von Reaktionskinetikrechnungen gewonnen. Hierfür wird zuerst die Gemischzusammensetzung in Abhängigkeit der Randbedingungen (Temperatur, Druck, Kraftstoffzusammensetzung, Luft-/Oxidatorzusammensetzung, Oxidator-Kraftstoffverhältnis, Abgasrückführrate, Wasserzusammensetzung und Wasserrate) bestimmt. Da die Abgaszusammensetzung im motorischen Betrieb immer von dem vorangegangenen Zyklus definiert ist, muss sie auch für diesen Anwendungsfall iterativ bestimmt werden. Zuerst wird für die Abgaszusammensetzung das Gleichgewichtsverhältnis des reinen Frischgases (ohne rückgeführtes Abgas) verwendet. Die Zusammensetzung für den nächsten Iterationsschritt kann dann aus den gesamten Gemischverhältnissen gewonnen werden. Dies geschieht so lange, bis die Differenz in den Speziesanteilen zwischen den einzelnen Iterationsschritten ein zuvor definiertes Minimum unterschreitet. Die Bestimmung der Abgaszusammensetzung erfolgt entweder weiterhin über die Gleichgewichtsbedingung (alle Reaktionen finden so lange statt, bis sich ein vollständiges Gleichgewicht einstellt) oder über einen zweiten Reaktor mit motorspezifischen Zeitskalen, nach deren Erreichen alle Reaktionen und Spezieskonzentrationen eingefroren werden. Ist die genaue Zusammensetzung des Gasgemisches bekannt, wird dann in der Reaktionskinetiksoftware Cantera eine Freiflamme über einem vordefinierten Netz von Stützstellen (in Flammenausbreitungsrichtung) aufgespannt. Für jede Stützstelle wird dann ein Differentialgleichungssystem über alle beteiligten Spezies und Reaktionsgleichungen gelöst. Als Ergebnis entsteht ein Flammenprofil mit den Verläufen von Temperatur, Flammengeschwindigkeit und Spezieskonzentrationen über der Flammendicke. In der Auswertung werden dann die laminare Flammengeschwindigkeit und die laminare Flammendicke (parallel über sieben verschiedene Definitionen) berechnet und für alle Stützstellen zusammengefasst.

Die analoge Berechnung der Zündverzugszeiten läuft sehr ähnlich ab. Für die Bestimmung der Gaszusammensetzung wird dieselbe Methodik wie für die laminaren Kenngrößen verwendet. Die darauffolgende Berechnung findet dann aber wahlweise in einem Konstantvolumen- oder in einem Konstantdruckreaktor statt. Das homogene Gemisch wird hier in absoluter Ruhe (ohne Turbulenzeinflüsse) mit den Randbedingungen (Temperatur, Druck,

Luft-Kraftstoffverhältnis und Abgasrückführrate) beaufschlagt. Der Status des Reaktors - und damit auch die Speziesanteile im Gemisch - wird dann für jeden Zeitschritt neu berechnet. Wird eine Maximalzeit überschritten oder eine eindeutige Selbstzündung erkannt, endet die Berechnung im Reaktor und die Datenaufbereitung für die aktuelle Stützstelle beginnt. Anhand mehrerer Kriterien wird hierbei die Hochtemperaturzündverzugszeit und - wenn zutreffend - die Niedertemperaturzündverzugszeit sowie der Temperaturanstieg durch die Niedertemperaturzündung bestimmt. Als Niedertemperaturzündung werden Vorreaktionen im Gemisch bezeichnet, die zu einer Temperatur- und Druckerhöhung führen, jedoch keine direkte Volumenzündung nach sich ziehen. Sind diese Werte für alle zuvor angeforderten Randbedingungen erlangt, so werden sie übersichtlich in einer Ausgabedatei ge-speichert. Durch die Implementierung von verteiltem Rechnen konnten diese beiden Prozesse (an der laminaren Freiflamme und im Reaktor) um mindestens eine Größenordnung (abhängig von der vorhandenen Rechenkapazität) beschleunigt werden.

Eine Analyse der für die Kraftstoffe erlangten Werte zeigte, dass - mit einer Ausnahme für Wasserstoff - alle existierenden Korrelationsgleichungen für die laminaren Flammengeschwindigkeiten und die laminaren Flammendicken weiterverwendet werden konnten. Für Wasserstoff wurde demnach der Entwurf einer angepassten Korrelation nötig. Bei dieser lag der Fokus auf bestmöglicher Robustheit und einfachem Aufbau für die automatisierte Abstimmung, ohne Qualitätseinbußen in Kauf nehmen zu müssen. Da die existierenden Korrelationsgleichungen für die Zündverzugszeiten nur für das Abbilden von verschiedenen Benzinqualitäten ausgelegt waren, mussten diese für alle betrachteten Kraftstoffe angepasst werden. Auch hier war wichtig zu gewährleisten, dass die neu geschaffene Korrelation sowohl das Verhalten von Benzinmischungen, als auch von (Synthetik-)Kraftstoffen oder deren Mischungen verlässlich abbilden kann.

Im automatisierten Prozess werden zuerst diejenigen Kraftstoffeigenschaften berechnet, die Reaktionskinetiksimulationen erfordern. Alle anderen Werte (spezifische Wärmekapazitäten, Dichten im flüssigen Aggregatszustand, thermische Leitfähigkeiten und kritische Punkte) werden dann durch dedizierte Mischungsregeln gewonnen. Die danach ablaufenden automatischen Korrelationsanpassungen für die laminaren Flammengeschwindigkeiten, laminaren Flammendicken und Zündverzugszeiten erlauben die schnelle und einfache Einführung von neuen Kraftstoffen und Kraftstoffmischungen, ohne dass der Prozess überwacht werden muss oder manuelle Eingriffe erforderlich sind. Für die Korrelation der laminaren Flammengeschwindigkeiten findet zuerst eine Abstimmung der Quereinflüsse von Temperatur und Druck bei stoichiometrischem Gemischverhältnis und ohne Abgasrückführung oder Wassereinspritzung statt. Hat diese Basiskonfiguration eine ausreichende Güte, so werden anhand einer für den späteren motorischen Betrieb signifikanten Temperatur- und Druckstelle nacheinander alle anderen Quereinflüsse variiert und angeglichen. Je nach geplantem Anwendungsfall können manche Einflüsse auch ausgeblendet und - falls benötigt - später nachgepflegt werden. Nach Beendigung aller Abstimmungsprozesse werden übersichtliche Fehlerdiagramme und die optimalen Variablengrößen ausgegeben. Letztere direkt in der Syntax, wie sie später für das Submodul in der Verbrennungssimulation benötigt wird. Dieses Vorgehen hält den manuellen Aufwand gering und hilft Übertragungs- und Tippfehler zu vermeiden.

Da die Arrhenius-Gleichungen in ihrer Grundform die Temperaturabhängigkeit bereits inkludieren, wird für die Basiskorrelation der Zündverzugszeiten nur der Einfluss von verändertem Umgebungsdruck abgeglichen. Dadurch, dass - im Gegensatz zu den laminaren Flammengeschwindkeiten - für alle sonstigen Einflüsse auf die Zündverzugszeiten noch Quereinflüsse über dem Druck auftreten, müssen diese ausreichend kompensiert werden. Dies geschieht, indem für alle weiteren Einflussgrößen einmal über die Größe selbst und dann noch einmal über den Druck iteriert wird. Für jede Kombination der zwei Einflüsse wird ein optimaler Wert gefunden. Diese Optima werden dann über kaskadierte Polynome angenähert. Für einen späteren Aufruf müssen damit nur noch die Polynomgleichungen und zugehörigen Koeffizienten gespeichert werden. Dies erlaubt schnelle Zugriffszeiten bei minimalem Speicherplatzbedarf.

Der Prozessablauf ist so gestaltet, dass für einen neuen Kraftstoff einmalig nur die Definition des Kraftstoffes selbst vom Nutzer angegeben werden muss. Handelt es sich hierbei um einen Misch- oder Surrogatkraftstoff (zum Beispiel TRF anstelle von Normalbenzin), so können die Speziesverhältnisse bei Standardbedingungen entweder über Mol-, Volumen- oder Massenverhältnisse angegeben werden. Außerdem muss ein Reaktionsmechanismus bereitgestellt werden, der für den gewünschten Kraftstoff oder die gewünschte Kraftstoffmischung geeignet und validiert ist. Nur so kann garantiert werden, dass plausible Rechenergebnisse erlangt werden können. Ist kein passender Reaktionsmechanismus vorhanden, so müssen alle Ergebnisse und damit auch die Optimierungsparameter der Gleichungen sehr bewusst interpretiert werden. Damit kann die Automatisierung des Prozesses zwar Fehler durch falsche Nutzung minimieren, dennoch sollte sie nur von Nutzern mit ausreichendem Hintergrundwissen in der Reaktionskinetik angewendet werden.

Um die neu vorgestellte, automatisierte Methode zu validieren, wurden Messdaten von sechs verschiedenen Motoren herangezogen. Diese unterscheiden sich in ihrem Anwendungsgebiet, ihrer Geometrie und den Betriebsbedingungen. Außerdem wurden zwölf verschiedene Kraftstoffe oder Kraftstoffmischungen herangezogen, um die Güte des Prozesses zu beurteilen: Normalbenzin, Cyclopentanon in wechselnden Mischungsverhältnissen mit Normalbenzin, Anisol in wechselnden Mischungsverhältnissen mit Normalbenzin, Methanol, ein Gemisch aus Dimethylcarbonat und Methylformiat sowie Wasserstoff. Zwei Verbrennungsmodelle (*Damköhler* und *Peters*) und zwei Klopfmodelle (*Fandakov* und *Hess*) wurden mit der neuen Methode getestet. Diese Vielzahl an Randbedingungen und Randbedingungskombinationen erlaubt es, eine sehr fundierte Aussage über die Güte und die Robustheit des Prozesses zu treffen. Für alle Motoren und Kraftstoffe zeigten sich sehr gute Ergebnisse für den Vergleich von Vohersage der Verbrennungsmodelle und Messung. Beim Vergleich der Klopfmodelle wurde sichtbar, dass die Methodik von *Fandakov* zwar die bessere Vorhersage liefert, das Modell von *Hess* nach einer Anpassung der Kalibrierung aber insgesamt eine bessere Qualität hat. Zusammenfassend zeigt sich, dass die vorgestellte Methodik allen an sie gestellten Anforderungen genügt. Sie kann für die Vorhersage der Verbrennung all der Kraftstoffe verwendet werden, für die ein ausreichend genauer Reaktionsmechanismus zur Verfügung steht. Die erreichbare Vorhersagequalität der betrachteten Modelle entspricht der bei der Simulation mit Normalbenzin.

1 Introduction

The deadline for the Paris climate goals is coming closer and closer but at the current date, the 1.5 °C limit seems impossible to achieve. While for new passenger cars in metropolitan areas, the electric powertrain seems to be the superior choice, ICEs will still have their applications. Replacing gasoline with CO_2 neutral fuels for the existing car fleets might be a niche case, but especially the transport sector and civil engineering are strongly dependent on high energy densities. Synthetic, carbon-neutral fuels can be ideal to decarbonize those areas. If new fuels then are created or existing engines are adapted to run with alternative fuels, it would be advantageous to profit from all benefits that come with it. Fuels with higher burn rates and greater knock resistance can achieve better thermodynamic efficiencies if the engine is properly designed for this kind of application. This can also counter the higher price of synthetic fuels to some extend.

To facilitate and accelerate the engine development for those new fuels, simulation is ideal. With it, the costly time on engine test benches can be drastically reduced and especially in the early design phases with lesser known fuels, the risk of damaging prototype components is mitigated. Particularly the Zero-Dimensional (0D) and Quasi-Dimensional (QD) simulation with its short calculation times allows for wide optimization runs and the prediction of transient engine and powertrain behavior. For this, all relevant fuel characteristics have to be integrated into the corresponding simulation models. While there are existing concepts for this, they usually require lots of manual work for the correlation fits and model adjustments. This work is aiming to get rid of the existing processes flaws and to create a complete, automated methodology for the predictive simulation of the combustion of single- and multicomponent fuels. Hereby, the automated process should deliver at least the same quality as it is the case with the current, manual procedure.

This work first introduces the needed physiochemical fundamentals and those of reaction kinetics calculations and engine combustion modeling. The methodology of all reaction kinetics calculations is shown and a method

S. K. Crönert, *A Complete Methodology for the Predictive Simulation of Novel, Single- and Multi-Component Fuel Combustion*, Wissenschaftliche Reihe Fahrzeugtechnik Universität Stuttgart, https://doi.org/10.1007/978-3-658-43075-7_1

of combining the python multiprocessing feature with Cantera objects and a timeout function is presented. Then, the automation process of a new fuel implementation is schematized. The automation begins with calculating all needed reaction kinetic data (laminar burning velocities, laminar flame thicknesses and ignition delay times), then proceeds with determining all mixture properties relevant for the engine combustion simulation. Afterwards, the correlations are automatically fit to the reaction kinetic calculation results. Then all simulation modules have got the information needed for predictive simulations of a fuel change. Before comparing the simulation results, the properties of the engines used and the fuel compositions are given. The whole methodology is validated on a wide range of engine designs and fuel/fuel mixtures by the means of burn rates. Additionally, for the two fuel mixtures of Anisole and Cyclopentanone with Gasoline, the predictive capabilities of the *Fandakov* and the *Hess* knock models are tested. Afterwards, a short summary and an outlook for things to improve conclude the work.

2 Fundamentals

Combustion is defined as the oxidization of a fuel in an exothermal reaction, usually with ambient air being used as the oxidizer. Flames - not necessarily visible - can form with *laminar* or *turbulent* surfaces, the former only under undisturbed conditions. While for some technical processes, combustion can be stable and uninterrupted over long periods of time (e.g., heating an industrial forge by burning methane with an open flame), this is not the case for internal combustion engines with intermittent combustion. A third property to characterize combustion is the distinction between *pre-mixed* and *diffusion controlled* flames.

All the research for this work is done on Spark Ignition (SI) engines (with direct injection). For QD simulation, this results in limiting the investigations to pre-mixed homogeneous flames. As the only fuel influences on turbulent combustion originate in the laminar flame, only laminar flame propagation is of interest for fuel specific investigations. The fundamentals needed to understand this type of combustion are treated in the following sections. For basic combustion engine fundamentals a consolidation of common literature is advised.

2.1 Physiochemical Fundamentals

2.1.1 Continuity Equations

The systems observed, both in reaction kinetic modeling and for the 0D/QD and One-Dimensional (1D) combustion and load exchange simulation, have to obey the fundamental laws of thermodynamics [113]. For the enclosed system, the conservation laws for overall mass, species masses, and overall enthalpy apply. At the flame front that marks the border between the two subsystems of *unburnt* and *burnt* mixture, mass and energy can be transported over the subsystem limits. As these exchanges have a significant influence

S. K. Crönert, *A Complete Methodology for the Predictive Simulation of Novel, Single- and Multi-Component Fuel Combustion*, Wissenschaftliche Reihe Fahrzeugtechnik Universität Stuttgart, https://doi.org/10.1007/978-3-658-43075-7_2

on the speed at which combustion occurs, they have to be considered for the simulations [110]. This section is integrated to allow for a basic understanding of the conservation laws, further information can be found in common thermodynamics textbooks (e.g., [29]).

A basic formulation of a conservation law is given in eq. 2.1. It consists of the differential examination of the conservation values density W_{c} with respect to time and its flux J with the location z. Jointly, both terms need to equal Q, denoting all sources or drains of the conservation value itself.

$$\frac{\partial W_{\mathrm{c}}}{\partial t} + \frac{\partial J}{\partial z} = Q \qquad \text{eq. 2.1}$$

Overall Mass Continuity

For the conservation of overall mass, equation eq. 2.1 is used for the density changes. As stated, there can be no mass transport over the system's border during combustion (only before and after, during cylinder load exchange), so the source term is 0 [113]. The density ρ is the density of the overall mixture. The density flow J is calculated by the velocity of the mixture's flux v weighted with its density. This results in equation eq. 2.2.

$$\frac{\partial \rho}{\partial t} + \frac{\partial (\rho \cdot v)}{\partial z} = 0 \qquad \text{eq. 2.2}$$

Species Masses Continuity

In contrast to the overall mass, the mass of chemical species can be changed by reactions [113]. The source term for each species results from reaction kinetics and is dependent on the molar mass M_{i} of species i and the change in species concentration (A) with time (see 2.2.2). This way, the source term r_{si} is given by equation eq. 2.3.

$$r_{\mathrm{si}} = M_{\mathrm{i}} \cdot \frac{\partial (\mathrm{A})}{\partial t} \qquad \text{eq. 2.3}$$

The partial species density ρ_{i} is calculated for each species by the following equation eq. 2.4, with w_{i} representing the mass fraction of species i. The same principle then also applies to the flux of the species masses in equation eq. 2.5.

$$\rho_i = \rho \cdot w_i \qquad \text{eq. 2.4}$$

$$J_i = \rho \cdot w_i \cdot v_i \qquad \text{eq. 2.5}$$

When the partial derivate of equation eq. 2.5 is formed, rearranged, and compared to equation eq. 2.2, this results in equation eq. 2.6. The first two terms correspond to equation eq. 2.2, the third term represents diffusion.

$$\rho \cdot \frac{\partial w_i}{\partial t} + \rho \cdot v \cdot \frac{\partial w_i}{\partial z} + \frac{\partial j_i}{\partial z} = r_{si} \qquad \text{eq. 2.6}$$

Enthalpy Continuity

In the case of reaction kinetic calculations, due to the simplification of adiabatic system boundaries and the fundamental validity of the first law of thermodynamics, there can be no source term for the overall enthalpy [113]. Depending on where the system limits are drawn in engine simulation, there can be sources of enthalpy flux (cylinder walls, intake/exhaust runner walls, etc.). But usually, wall heat transfers are modeled. The overall enthalpy H is equal to the sum of all species enthalpies h_i weighted by the correlating species density ρ_i as shown in equation eq. 2.7. The flux term is defined by the species' movements, which are superimposed by the turbulence of the overall mixture. Additionally, it is influenced by heat transfers (see 2.1.3). Equation eq. 2.8 shows those correlations, equation eq. 2.9 is the resulting formula for the continuity of the overall enthalpy [52].

$$H = \sum_i \rho_i \cdot h_i \qquad \text{eq. 2.7}$$

$$J = \sum_i \rho_i \cdot v_i \cdot h_i + \phi_q \qquad \text{eq. 2.8}$$

$$\sum_{\mathrm{i}} \frac{\partial(\rho_\mathrm{i} \cdot h_\mathrm{i})}{\partial t} + \sum_{\mathrm{i}} \frac{\partial(\rho_\mathrm{i} \cdot v_\mathrm{i} \cdot h_\mathrm{i})}{\partial z} + \frac{\partial \phi_\mathrm{q}}{\partial z} = 0 \qquad \text{eq. 2.9}$$

2.1.2 Species Transport

The transport of fuel and oxidizer into the flame front and the removal of products away from it are highly influential on the speed of combustion [113]. This is based on the Chatelier's principle (see 2.2.4), which states that a decrease in product concentrations and an increase in the reactant concentrations will overall lead to an accelerated reaction. The diffusion flux that is caused by concentration differences in a mixture for species i in a mixture of j species is given by equation eq. 2.10.

$$j_\mathrm{i} = \frac{c_\mathrm{i}^2}{\rho} \cdot M_\mathrm{i} \cdot \sum_{j} (M_j \cdot D_{\mathrm{i}j} \cdot \frac{\partial x_j}{\partial z}) \qquad \text{eq. 2.10}$$

Where c_i is the molar concentration of species i, ρ the density, M_i and M_j the molar masses of the corresponding species. The sum also takes the diffusion coefficient for the diffusion from species i into j $D_{\mathrm{i}j}$ and the molar distribution of species j into account. It can be interpreted as the entirety of all individual diffusion streams of the species i into all other species $j \in \{1, \ldots, n\} \backslash \{\mathrm{i}\}$ in the mixture. This also means that the transport equation eq. 2.10 has to be modelled for each species individually. If a mixture-averaged, multi-component diffusion coefficient D_mix is used, the calculation can be drastically simplified and sped up. This is reasonable, especially when one of the mixture species concentrations is dominating, just as it is the case with liquid fuel combustion in atmospheric air [113].

2.1.3 Heat Transfer

The phenomenon of the transport of inner energy from one body (or fluid) to another is called heat transfer. It can be further specified into three sub-mechanisms. Thermal conduction describes the transfer of heat by contact between two systems. This is caused by direct transmission at the molecular level. The exchange by thermal convection occurs when a fluid carries its inner energy into a second one, most of the time accompanied by diffusion. The effect of transporting energy by the means of electro-magnetic waves is called thermal radiation [18][89]. Thermal convection is already compensated for when obeying the conservation of the overall enthalpy, so it will not be covered further (see: 2.1.1). Because of its low absolute heat transfers, heat radiation does only have a minor impact on the flame propagation and engine-relevant processes, nevertheless it is mentioned for the sake of completeness. For interested readers, a further study of the literature in [3][46][57] and [89] is recommended.

Thermal Conduction

Thermal conduction causes the warming of the pre-heating zone. This way, the activation energy for any pre-reactions in the unburnt mixture is provided. The higher the thermal conduction coefficient is, the higher the flame propagation speed will be. Once the thermal conductivity is too high, growing heat exchange with the system limits - for ICEs especially with the cylinder walls can cause opposite effects, up to an early quenching of the flame. Fourier's law of thermal conduction defines the heat flux ϕ_{q} as the product of the mixture's specific thermal conductivity coefficient λ and the gradient of the local temperature T [113]. It is shown in equation eq. 2.11.

$$\phi_{\mathrm{q}} = -\lambda \nabla T \qquad \text{eq. 2.11}$$

The thermal conduction is highly influenced by the specific heat capacities of the mixture components. This especially shows up in variations of the fuel-air-mixture, with higher exhaust gas recirculation ratios and with externally injected water. Up to a specific point or depending on the operating point of the engine, this is desired and helps to counteract knocking behavior or the production of nitrogen oxydes. For isolated mixtures and species, the intrinsic property of the thermal conductivity coefficient can be calculated by imposing a known heat flux onto them. For mixtures containing known amounts of multiple species, the mixture's thermal conductivity coefficient is calculated by the mixing rule in equation eq. 2.12.

$$\lambda_{\mathrm{mix}} = 0.5 \left(\sum_{\mathrm{i}} x_{\mathrm{i}} \lambda_{\mathrm{i}} + \frac{1}{\sum_{\mathrm{i}} x_{\mathrm{i}} / \lambda_{\mathrm{i}}} \right) \qquad \text{eq. 2.12}$$

Thermal Radiation

Heat transport from one body or fluid to another by the means of electro-magnetic radiation is called thermal radiation [46]. Every molecule of substance whose temperature is above absolute zero emits such radiation. For dense liquids or solid bodies, only the surface molecules can emit radiation to their surroundings, the internal molecules transport their energy to adjacent ones. Because of this, solid bodies are also called *surface emitters*, gases *volume emitters*. The highest amount of thermal radiation for combustive processes is caused by molecules of the species CO_2 and H_2O. Compared to N_2, their concentrations in the burnt gas mixture are quite low, which limits the heat exchange into the fresh mixture [18]. This is why thermal radiation is ignored for combustion simulation and not modelled any further.

Specific Heat Capacities, Enthalpies, and Entropies of Fuel Mixtures

For engine simulation, the specific heat capacities of all involved compounds are important for an accurate determination of mixture temperatures. As the fuel concentration compared to fresh air and recirculated exhaust gases is usually rather low, a deviation in its heat capacity does not have a big impact on overall cylinder temperatures. Nevertheless, proper modeling is advised. This work implies that the specific heat capacities (better: their polynomial coefficients) of all species involved are known. If this is not the case, *Grill et al.* showed, that for most common fuels, their caloric properties can be estimated by an approximation based on the H/C ratio [44].

The mixture's specific heat capacity for constant pressures is then calculated according to equation eq. 2.13 by summing up the individual species' heat capacities c_{pi}, weighted by their molecular fractions x_{i}. When this is done for the germane temperature range, the polynomial in equation eq. 2.16 can be fitted. It is noted that for a proper fit, usually two polynomials for two temperature ranges are used. The low temperature polynomial is fitted to all values for temperatures below or equal to 1000 K, and the high temperature polynomial for everything above.

$$c_{\mathrm{p,mix}} = \sum_{\mathrm{i}} x_{\mathrm{i}} c_{\mathrm{p,i}} \qquad \text{eq. 2.13}$$

For the temperature dependence of the mixture's enthalpy and entropy, the same mixing rules apply. They are calculated analogue to the specific heat capacity by equations eq. 2.14, and eq. 2.15.

$$H_{\mathrm{mix}} = \sum_{\mathrm{i}} x_{\mathrm{i}} H_{\mathrm{i}} \qquad \text{eq. 2.14}$$

$$S_{\mathrm{mix}} = \sum_{\mathrm{i}} x_{\mathrm{i}} S_{\mathrm{i}} \qquad \text{eq. 2.15}$$

As the main influence on all those fuel properties for engine specific operating conditions lies in temperature changes, only their temperature dependence is modelled. The most common method of doing this is fitting measurement values derived from caloric experiments to higher order polynomials[1]. Those polynomials are called NASA-polynomials after their origin and exist in two ways [1]: The old, 7-term version has been in use for decades and additionally a newer, 9-term model has been introduced to improve the fit quality. Since the newer version is used in this work, only its equations are presented. The heat capacity is modelled by equation eq. 2.16, the enthalpy by equation eq. 2.17, and the fuels' absolute entropy by equation eq. 2.18. They all share the same set of polynomial coefficients and have to be fitted simultaneously.

$$\frac{c_{\mathrm{p}}}{R} = a_1 \frac{1}{T^2} + a_2 \frac{1}{T} + a_3 + a_4 T + a_5 T^2 + a_6 T^3 + a_7 T^4 \qquad \text{eq. 2.16}$$

$$\frac{H}{RT} = -a_1 \frac{1}{T^2} + a_2 \frac{\ln T}{T} + a_3 + a_4 \frac{T}{2} + a_5 \frac{T^2}{3} + a_6 \frac{T^3}{4} + a_7 \frac{T^4}{5} + \frac{b_1}{T} \qquad \text{eq. 2.17}$$

[1] This is - apart from the coefficient count that might differ - exactly the same way as it is done in the chemical reaction kinetics mechanisms, see 2.2.1.

$$\frac{S}{R} = -a_1 \frac{1}{2T^2} - a_2 \frac{1}{T} + a_3 \ln T + a_4 T + a_5 \frac{T^2}{2} + a_6 \frac{T^3}{3} + a_7 \frac{T^4}{4} + b_2 \quad \text{eq. 2.18}$$

The NASA polynomials are used for the fuels and fuel mixtures in this work are provided in Appendix A.

2.1.4 Other Intrinsic Mixture Information

Atom Counts

The ratio of the fuel mixture atom counts is needed for the determination of the stoichiometric Air-to-Fuel Ratio (AFR) and emission modeling. It is calculated by summing up the atom counts of all mixture components weighted by their species share following eq. 2.19.

$$n_{\mathrm{C,\,mix}} = \sum_{\mathrm{i}} x_{\mathrm{i}} n_{C,\mathrm{i}} \quad \text{eq. 2.19}$$

Lower Heating Value

The lower heating value is needed for engine simulation as its product with the fresh mixture mass conversion rate results in the burn rate (see 2.3.1). As it can be determined using the NASA polynomial for the enthalpy, there is no need to calculate it independently. It is defined as the enthalpy of the fuel mixture minus the enthalpy of its complete combustion products, both evaluated at the standard temperature of 298.15 K.

$$LHV = H_{\mathrm{fuel,\,298.15\,K}} - H_{\mathrm{comb,\,298.15\,K}} \quad \text{eq. 2.20}$$

Absolute Entropy at Standard Temperature and Pressure

As well as the lower heating value, the absolute entropy of a mixture at standard temperature and pressure can be calculated by the aforementioned NASA polynomials. Since no pressure dependence is given, it is enough to solve eq. 2.18 for T = 298.15 K.

Dynamic Viscosities

For the calculation of the mixture's dynamic viscosities, a model from *Kee et al.* [61] that is based on *Wilke* [115] mixing rules is used. The relation between the dynamic viscosity of the mixture η_{mix} and those of the isolated species is drawn in eq. 2.21. The symbol ϕ is a dimensionless constant further specified in eq. 2.22.

$$\eta_{\mathrm{mix}} = \sum_{\mathrm{i}} \frac{\eta_{\mathrm{i}} x_{\mathrm{i}}}{\sum_j \phi_{\mathrm{i}j} x_j} \qquad \text{eq. 2.21}$$

$$\phi_{\mathrm{i}j} = \frac{\left[1 + \sqrt[2]{\frac{\eta_{\mathrm{i}}}{\eta_j}} \sqrt[4]{\frac{M_j}{M_{\mathrm{i}}}}\right]^2}{\frac{4}{\sqrt[2]{2}} \sqrt[2]{1 + \frac{M_{\mathrm{i}}}{M_j}}} \qquad \text{eq. 2.22}$$

Critical Point

Multiple methods have been proposed to reliably calculate the critical point of a mixture based on the properties of its components [51] [74] [83] [108]. For this work, the critical point of a binary mixture has been calculated based on its pure component properties following the equations proposed by *Lee and Kesler* [72], later reworked by *Pedersen et al.* [91]: The critical volume V_{c} is calculated by eq. 2.25, the critical temperature T_{c} by eq. 2.26, and the critical pressure p_{c} by eq. 2.27.

$$V_{\mathrm{c}ij} = \frac{1}{8}\left(V_{\mathrm{c,i}}^{\frac{1}{3}} + V_{\mathrm{c},j}^{\frac{1}{3}} \right)^3 \qquad \text{eq. 2.23}$$

$$T_{\mathrm{c}ij} = (T_{\mathrm{ci}} T_{\mathrm{c}j})^{1/2} \qquad \text{eq. 2.24}$$

$$V_{\mathrm{c,mix}} = \sum_{\mathrm{i}}^{n} \sum_{j}^{n} x_{\mathrm{i}} x_j V_{\mathrm{c i}j} \qquad \text{eq. 2.25}$$

$$T_{\mathrm{c,mix}} = \frac{\sum_{\mathrm{i}}^{n} \sum_{j}^{n} x_{\mathrm{i}} x_j V_{\mathrm{c i}j} T_{\mathrm{c i}j}}{\sum_{\mathrm{i}}^{n} \sum_{j}^{n} x_{\mathrm{i}} x_j V_{\mathrm{c i}j}} \qquad \text{eq. 2.26}$$

$$p_{\mathrm{c,mix}} = \frac{R \cdot \overline{Z_{\mathrm{c}}} T_{\mathrm{c,mix}}}{V_{\mathrm{c,mix}}} \qquad \text{eq. 2.27}$$

$$with \qquad \overline{Z_{\mathrm{c}}} = \sum_{i}^{n} x_{\mathrm{i}} Z_{\mathrm{c,i}}$$

Liquid Density

The simple mixing rules as in eq. 2.13 are used for the calculation of liquid densities of mixtures. As they are usually given in gravimetric relations, the equation is rearranged to eq. 2.28.

$$\frac{1}{\rho_{\mathrm{liq,mix}}} = \sum_{\mathrm{i}} \frac{w_{\mathrm{i}}}{\rho_{\mathrm{liq,i}}} \qquad \text{eq. 2.28}$$

2.2 Fundamentals of Reaction Kinetics

While (chemical) thermodynamic equations are able to model the outcome of an arbitrary physicochemical process (e.g., adiabatic flame temperature of a CH_4/O_2-mixture), they provide no information about the speed with which underlying reactions take place [111] [113]. For these purposes, reaction kinetic calculations are used. This way, it is possible to describe time-resolved processes of chemical reactions.

Depending on the problem to be solved with reaction kinetics, the results in either time (ignition delay) or the linear flame dimension (laminar flame speed and laminar flame thickness) are of interest. A grid is created to extend across this dimension, and then for all grid points, a system of differential equations has to be solved with each of the equations being represented by a single elementary reaction and all of them sharing the same pool of chemical species.

The most important input for those reaction kinetic calculations are the reaction mechanisms (see 2.2.1). They include all the physical (transport properties, thermal capacities) and chemical (elementary reactions, reaction rates) information needed to solve the differential equations.

2.2.1 Reaction Mechanisms

To gather information for gross reactions outside of feasible boundary conditions, simple measurements and an *Arrhenius*-fit (as in eq. 2.35) are not possible any more. The process of the reaction has to be modelled by breaking it down into multiple elementary reactions (see 2.2.2). The combination of several of these takes place in a so-called reaction mechanism. It does not only contain all the major elementary reactions with their corresponding reaction rates, but also the thermophysical species properties needed for a complete depiction of the chemical system. This way, reaction mechanisms fulfill all chemical and physical fundamentals and allow for an extrapolation into regimes where no measurement data for validation (e.g., laminar flame speed or ignition delay times) is available. Mechanisms that include all elementary reactions of a gross reaction are called *complete*.

The relationship between multiple elementary reactions is called coupling. It is divided into *parallel* and *consecutive* coupling [65]. To depict this idea, simple examples of unimolecular reactions are used (examples also taken from [65]), but the same effects apply to bimolecular reactions as well. For the relationship between two reactions, both having reactant species A but different products (species B and species C), the parallel coupling follows the scheme in eq. 2.29. As both elementary reactions share the same reactants, changes in their equilibrium will also affect the corresponding reaction (see 2.2.4).

$$A \begin{array}{l} \nearrow B \\ \searrow C \end{array} \qquad \text{eq. 2.29}$$

Consecutive reactions follow the scheme in eq. 2.30. Here, the coupling between the two corresponding elementary reactions is created by the shared species B. When the product concentration for the first reaction increases, so does the reactant concentration for the second reaction. Species B is called an *intermediate* species.

$$A \longrightarrow B \longrightarrow C \qquad \text{eq. 2.30}$$

The higher the complexity of the reaction (not necessarily apparent by the look of a sum formula), the greater the count of elementary reactions in a mechanism [113]. This also influences the number of links (*parallel* and *consecutive*) between them. For simple reactions, e.g., the combustion of methane with air, all elementary reactions are known and well studied. For more complex processes, including the oxidation of long-chain hydrocarbons, the information available is more sparse. But even for the supposedly simple combustion of small molecules, a complete description may require a decent number of elementary reactions (roughly 400 for the combustion of methane in air). Since not all reactions show the same sensitivity for some calculations, they can be compressed if applicable. With a sensitivity analysis, the reactions with the highest impact on the gross reaction can be identified and isolated. With this, the number of differential equations to solve and, with it, the time for the calculations can be significantly decreased. This means, depending on the use case, smaller reactions can be used to improve efficiency. Since usually not every influence of all elementary reactions is exactly known, even complete mechanisms have to be calibrated to measurement data. This also means that, depending on the validation process and application scenario, some mechanisms meet the requirements better than others. A more complex mechanism does not necessarily yield better results for simple calculations (as the aforementioned methane combustion) than one that has been specifically designed for this function. Even with the best post-processing and great calibration fits, results calculated with an improper reaction mechanism will not be of any use.

All reaction mechanisms have the same onion-like internal structure; an example is depicted in figure 2.1. The innermost circle always contains the most basic form of oxidation. It participates in the combustion of any fuel containing hydrogen. If any (hydro-)carbon components are present, the second stage (CO oxidation) is added. All inner circles are also needed for more complex reactions of long-chain hydrocarbon fuels as they are decomposed into smaller intermediate products prior to complete conversion. Because of this structure, the more complex a fuel is, the more complex the mechanism to describe the fuel usually is [113].

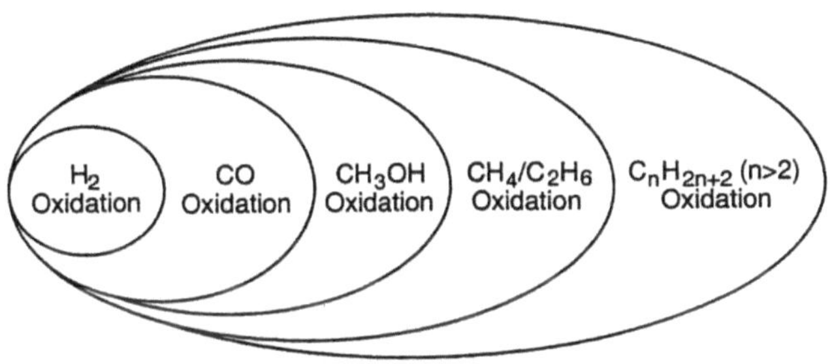

Figure 2.1: Onion-like integration of sub-mechanisms in more complex reaction mechanisms [113].

As already stated, the choice of the right reaction mechanism is important for the calculation results. It has to be designed specifically for the right fuel and application. While a more complex mechanism should include all the needed reactions to calculate the laminar flame propagation of methane, there is no guarantee that the reaction rates that have been tweaked to fit the intended use case are not interfering. Additionally, a reaction mechanism designed for the calculation of laminar free flames is not feasible for ignition delay calculations, as the low temperature chemistry is absent. Vice versa, the results may be usable, but the calculation times needed to solve the system of differential equations for a laminar flame with complete low temperature chemistry are significant.

2.2.2 Elementary Reactions

As previously stated, elementary reactions are the attempt to break down complex chemical processes (for simplicity reasons in this case the oxidization of methane in pure oxygen) into the most basic sub-reactions [113]. When looking at oxidization of methane in eq. 2.31, it can be seen, that even for simpler reactions, the collision theory of *Traut* and *Lewis* can not accurately describe the processes occurring [60]. A single collision would have to break up all four C−H bindings at once [57]. In reality, the connections between the carbon and the hydrogen atoms will be split one by one, introducing intermediates like CH_3 and OH to the equation.

$$CH_4 + O_2 \rightleftarrows CO_2 + H_2O \qquad \text{eq. 2.31}$$

To address this, elementary reactions are introduced. Typically, these are represented by collisions of two molecules, with the kinetic energy of the reactants (or of another molecule that does not get changed in the process) providing the activation potential E_a - that form one or multiple reaction products. While theoretically those reactions are possible even with three reactants, it has been shown by *Cook et al.* that those cubic reactions can be expressed as a combination of two elementary reactions with up to two reactants [21]. One of the most common elementary reactions, the one of hydrogen in plain oxygen into oxygen and a hydroxy radical, is shown in eq. 2.32.

$$H + O_2 \rightleftarrows O + {}^{\bullet}OH \qquad \text{eq. 2.32}$$

It follows the basic scheme of a second order reaction as shown in eq. 2.33. A and B are called reactants, C and D products. The arrows in between them show that forward and reverse reactions can occur[2].

[2] To make aware that - in the internal combustion engine time scales - chemical reactions usually do not reach their equilibrium (even if it exists), all reactions in this work are written with the symbol for reversible reactions $\rightleftarrows$ rather than the one for chemical equilibrium $\rightleftharpoons$.

$$A + B \underset{\kappa_r}{\overset{\kappa_f}{\rightleftharpoons}} C + D \qquad \text{eq. 2.33}$$

The speed at which reactants will convert into products is defined as the reaction rate RR_f. It depends on the concentrations of all species involved, as well as on temperature and pressure. It can be seen as the change in concentration of a single species over time [39]. As chemical equations are read from left to right, positive reaction rates mean an increase in product and a decrease in reactant concentrations. Thus, the signing of reaction rate and the rate of change in concentration for an reactant has to be opposite, depicted in eq. 2.34. If the forward and reverse reaction rates are equal, no changes in species concentrations can be noticed any more, the equation or the system has reached is equilibrium (see 2.2.4). Reactions can still occur, but the production and use of all species is equal.

$$RR_f = \kappa_f(A)(B) = \frac{\partial n_f}{\partial t} = -\frac{\partial n_r}{\partial t} = -\frac{\partial(A)}{\partial t} \qquad \text{eq. 2.34}$$

Arrhenius modeled the forward reaction rate constant κ_f and its dependence on the activation energy and temperature according to eq. 2.35 [2]. Where A is referring to the collision frequency of the species involved, $e^{\frac{-E_a}{RT}}$ to the Boltzmann factor [8]. The latter is the fraction of all collisions with an energy exceeding E_a.

$$\kappa_f = Ae^{\frac{-E_a}{RT}} \qquad \text{eq. 2.35}$$

It was later extended by adding a temperature dependence of the pre-exponential factor A and has then be referred to as the *modified* Arrhenius equation [69]. The values for A, b, and E_a are specific to each reaction and have to be gathered experimentally. They are usually tabulated in the reaction mechanism. As there is no direct way of measuring them, they are adjusted to fit simulation results to experimental data (ignition delay or laminar flame speed measurement results).

$$\kappa_f = AT^b e^{\frac{-E_a}{RT}} \qquad \text{eq. 2.36}$$

These observations are true for all chemical reactions except recombination reactions and free-radical reactions with low activation energy. In both of these exemptions, the temperature dependence in the pre-exponential term has an increased influence [39].

2.2.3 Combustion Intermediates and Radicals

Another possibility to classify reactions in the combustion theory is the amount of radical species consumed or created in their process [110]. They are then divided into *chain-initiating* reactions, *chain-propagating* reactions, *chain-branching* reaction, and *chain-terminating* reactions. An overview for the exemplary reaction of hydrogen with oxygen can be found in table 2.1. *Chain-initiating* reactions are initially producing radicals that can take part in following reactions, in this case its in the form hydroxy radicals. *Chain-propagating* and *chain-branching* reactions are quite similar as they both either keep the radical count unchanged or increase it. The latter even speed the gross reaction up and are the reactions most responsible for detonations and flame propagations. With *chain-terminating* reactions, radical species are consumed without creating any of them as products. If the termination reactions are more dominant than *chain-initiating* and *chain-branching* reactions, the combustion will halt or not start at all. This might be the case for certain equivalence ratios or boundary conditions.

Table 2.1: Exemplary review of chain type of reaction for the elementary reactions (excerpt) of hydrogen oxidation [113].

H_2	+	O_2	=	$2\,OH^\bullet$			chain-initiating
$OH^\bullet$	+	H_2	=	H_2O	+	$H^\bullet$	chain-propagating
$H^\bullet$	+	O_2	=	$OH^\bullet$	+	$O^{\bullet\bullet}$	chain-branching
$H^\bullet$	+	$H^\bullet$	=	H_2			chain-terminating

2.2.4 Chemical Equilibrium

If forward and reverse reactions are occurring at the same speeds, the state of the reaction is called chemical equilibrium. While local (at the elementary reaction level) and global (at the gross reaction level) equilibria can and do exist in reaction chemistry, in the process of reaching them, the reactions are influenced by external interferences [67]. If the boundary conditions change, the reaction will adjust to counteract this external force. This effect is named after its discoverers, *Le Châtelier* and *Braun.*

According to them, there are three external influences that reactions encounter:

- **Temperature changes**
  An increasing temperature will speed up the reaction rate of the endothermal path, while a decreasing temperature will increase the one of the exothermal path. The previously mentioned exothermal oxidation of hydrogen with oxygen will prefer the forward reaction to O and $^{\bullet}OH$, if the system temperature is lowered.

$$H + O_2 \rightleftharpoons O + {}^{\bullet}OH \qquad \text{eq. 2.37}$$

- **Pressure changes**
  A compression of the system will increase the partial pressure of all species. The equilibrium will shift to the side with the smaller volume. With ideal gases, all molecules are of the same volume, and the reaction will promote the direction that decreases the overall mole count. For the *Haber & Bosch* ammonia synthesis, this results in more ammonium production (half the mole count). For a pressure decrease, inverse effects will apply.

$$N_2 + 3\,H_2 \rightleftharpoons 2\,NH_3 \qquad \text{eq. 2.38}$$

- **Concentration changes**
  If any of the species is drawn from the system, the reaction will increase the production (or limit the consumption) of this species. For the aforementioned ammonia synthesis, this is provoked by cooling down the mixture, even further promoting ammonia production (exothermal reaction). The

ammonia condenses at significantly higher temperatures than the reactants and can be harvested.

For the interpretation of kinetic effects, it is important to keep these phenomena in mind, as all elementary reactions are constantly contesting the same species. Plus, exo- or endothermal reactions are promoting or demoting certain reaction paths. For combustion processes, some reactions will never reach equilibrium, either because they are occurring too slowly for combustive time scales or because the reactive species (see radicals, 2.2.3) are already consumed by competing reactions. They then are not able to reach their equilibrium and are treated as *frozen*.

2.2.5 Laminar Flames

Even if no simple laminar combustion occurs in ICEs, it is still based on the same laminar base structure. Thus, to understand and model turbulent premixed combustion, it is necessary to first grasp the idea of laminar (premixed) flames. Since all (common) turbulent flame speed models are based on scaling laminar flame parameters (laminar burning velocity, laminar flame thickness) with turbulence, this idea is further reinforced. A flame is a "self-sustaining propagation of a localized combustion zone at subsonic velocities" [110]. This means, it is propagating through the unburnt mixture, converting reactant to intermediate and then final combustion products on its way. Laminar flames propagate only in one direction, normal to their flame front [94].

Laminar flames are divided into three zones (also see figure 2.2): The preheating zone, where energy from the flame front is conducted into the uburnt regime. Due to concentration differences, the fresh mixture reactants diffuse into the flame while combustion intermediates (radicals) get into the area in front of the flame. The reaction zone, where the main reactions, and consequently the major part of the heat release, take place. And the oxidation zone, where intermediate combustion products are oxidized and some additional temperature rise is achieved [68].

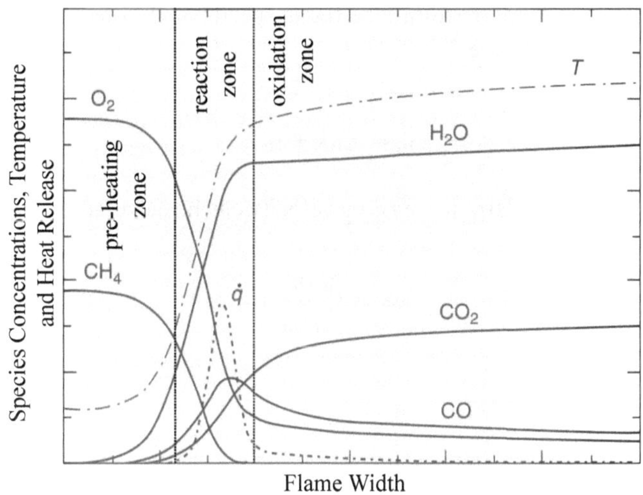

Figure 2.2: Flame profile (simulated) of a laminar flame with species concentrations, temperature profile, and heat release. The flame is divided into a pre-heating, a reaction, and an oxidation zone.

Laminar Burning Velocities

While there are multiple methods of measuring Laminar Burning Velocitys (LBVs) (explosion bombs, *Bunsen* burners, and heat flux burners), only the former will be described here since it is the most commonly used. Note that no own LBVs measurements are conducted as part of this work, by consequence the LBVs values used from other sources might be based on alternative methods. In an explosion bomb, a steady, pre-conditioned (pressure, temperature, composition) fuel-air-mixture in a spherical vessel is ignited with a pre-defined ignition energy [71] [88]. *Chen et al.* [19] showed, for the example of hydrogen/air-mixtures, that the registered flame propagation in a quasi-isobaric state is spherical and laminar. The local pressure or density differences due to the propagating flame front can be registered optically with *Schlieren*-Photography [105]. Post-processing of this data delivers the flame speed, including stretch effects. To correct these, the linear correlation between non-stretched and stretched flames with the *Markstein* length of the burnt gas $\mathcal{L}_b$ is used (eq. 2.39) [20] [117]. With this, linear or non-linear extrapolations are used to acquire the flame speeds of regions without flame stretch, as shown in figure 2.3.

$$s_{\mathrm{b}}^{0} = \mathcal{L}_b k + s_{\mathrm{b}} \qquad \text{eq. 2.39}$$

Once s_{b}^{0} is known, the laminar burning velocity can be calculated by eq. 2.40.

$$s_{\mathrm{L}} = s_{\mathrm{b}}^{0} \frac{\rho_{\mathrm{b}}}{\rho_{\mathrm{ub}}} \qquad \text{eq. 2.40}$$

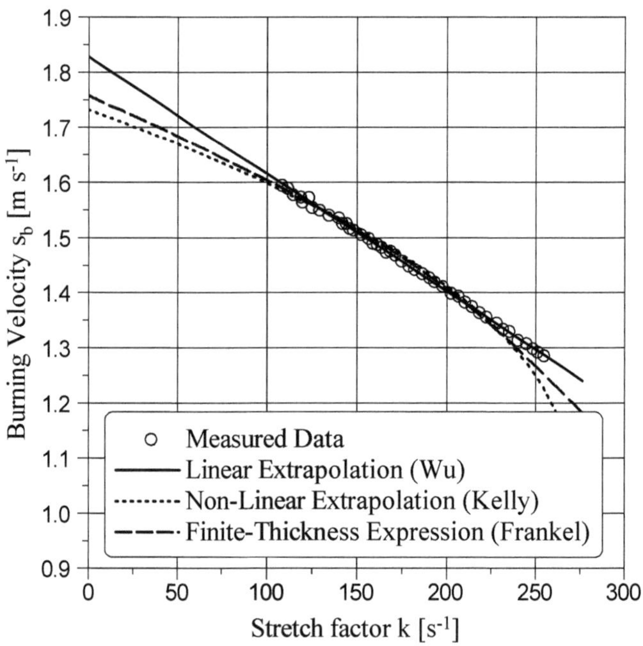

Figure 2.3: Extrapolation of burning velocity measurement data to the velocity of an unstretched flame. Data provided by [75]. The extrapolation methods are referring to *Wu et al.* (LE) [117], *Kelley* and *Law* (NE) [63], and *Frankel et al.* (FTE) [38].

Hydrodynamic instabilities and auto-ignition are limiting the area of boundary conditions where measured laminar burning velocities can be obtained to low pressures and rather low temperatures. Exemplary measurement data for gasoline, gasoline/ethanol mixtures, methane, and natural gas has been collected for figure 2.4. For combustion engine relevant data, extrapolations or simulation models need to be used (see *Laminar Burning Velocity Correlations* 2.3.3 and *Laminar Burning Velocity Calculations* 3.2).

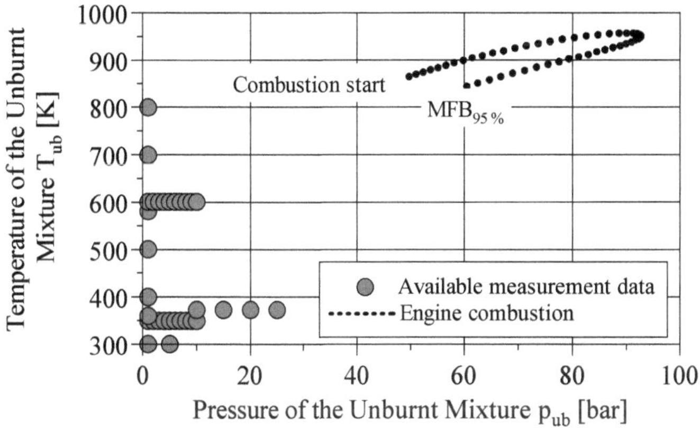

Figure 2.4: Measured laminar burning velocities available in literature (for gasoline and natural gas) in comparison to engine relevant pressures and temperatures. Figure adapted from [48], measurement data originally taken from [9], [14], [27], [28], [56], [59], [112].

Laminar Flame Thickness

The exact definition of the laminar flame thickness is difficult as the flame's limits cannot be exactly drawn. This circumstance also makes every attempt in defining it somewhat arbitrary. To gather measurement data for laminar flame thicknesses, usually the temperature profile of the flame is captured with optical methods. Depending on the data available, different sources recommend multiple ways of calculating the laminar flame thickness ([42] [93] [99]). Equations (eq. 2.43a) to (eq. 2.43h) show the methods that have been implemented into the calculation script, most of them based on the formulation in eq. 2.41, only varying the definition of the location of the inner layer [42]. The two remaining definitions rely on processing the flame temperature profile (*Ewald*, *therm*) similar to eq. 2.42 [99]. All of the used definitions are itemized in the following enumeration, the relative and absolute differences of their calculation results can be seen in figure 2.5.

$$\delta_{\mathrm{L}} = \frac{\lambda_{\mathrm{ub}}}{\rho_{\mathrm{ub}} c_{\mathrm{p,ub}}} \frac{1}{s_{\mathrm{L}}} \qquad \text{eq. 2.41}$$

$$\delta_{\mathrm{L}dT} = \frac{T_{\mathrm{ad}} - T_{\mathrm{ub}}}{max(\frac{\partial T}{\partial z})} \qquad \text{eq. 2.42}$$

(a) **Ewald**: Temperature gradient at the point of the highest Heat Release Rate (HRR) [33].

(b) ∇T: Temperature delta divided by the maximum of the temperature gradient.

(c) $\mathbf{H_{2max}}$: At the location of the maximum H_2 concentration [93].

(d) $\boldsymbol{\nu_{H2}}$: Ratio between kinematic viscosity of hydrogen ν_{H2} and the laminar burning velocity s_{L} [93].

(e) $\mathbf{(H_2 + CO)_{max}}$: At the location of the maximum of the sum of H_2 and CO concentrations [97].

(f) $HRR_{\mathbf{max}}$: At the location of the highest heat release rate [93].

(g) $\mathbf{{}^{\bullet}OH_{max}}$: At the location of the highest $^{\bullet}OH$ concentration [97].

(h) **therm**: Distance between where the flame temperature is 5 % higher than the minimum temperature and where the flame temperature is 5 % lower than the maximum temperature.

The comparison of those formulations in figure 2.5 shows that, apart from a few outliers, they deliver almost the same relative dependencies of the equivalence ratio. As this is the same for all fuels treated in this work, it is important to be consistent with the definition. The only additional limitation is that the two definitions that include the H_2 concentration are obviously unsuitable for any combustion with fuel containing hydrogen. Because this work is closely connected to that of *Hann*[48], it has to be noted that a definition different from the one used there has been chosen. This was done to ensure compatibility with a wider variety of fuels and fuel mixtures. From here on, whenever the laminar flame thickness δ_{L} is used, this refers to the one with the location of the inner layer at the maximum heat release rate (eq. 2.43f). Since the calculation time needed for post-processing the laminar flame data is marginal, all definitions are calculated simultaneously and stored for possible later use.

$$\delta_{\mathrm{L,Ewald}} = \frac{T_{\mathrm{end}} - T_{\mathrm{start}}}{\nabla T\big|_{\mathrm{HRR,\ max}\pm 5}} \qquad \text{eq. 2.43a}$$

$$\delta_{\mathrm{L},\nabla T} = \frac{T_{\mathrm{max}} - T_{\mathrm{start}}}{\nabla T_{\mathrm{max}}} \qquad \text{eq. 2.43b}$$

$$\delta_{\mathrm{L,H_2,max}} = \frac{\lambda_{\mathrm{th}}\big|_{\mathrm{H_{2max}}}}{c_{\mathrm{p}}\big|_{\mathrm{H_{2max}}}\rho\big|_{\mathrm{H_{2max}}}}\frac{1}{s_{\mathrm{L}}} \qquad \text{eq. 2.43c}$$

$$\delta_{\mathrm{L,H_2kin}} = \frac{\nu_{\mathrm{H_2,\ max}}}{s_{\mathrm{L}}} \qquad \text{eq. 2.43d}$$

$$\delta_{\mathrm{L,H_2CO,max}} = \frac{\lambda_{\mathrm{th}}\big|_{(\mathrm{H_2+CO})_{\mathrm{max}}}}{c_{\mathrm{p}}\big|_{(\mathrm{H_2+CO})_{\mathrm{max}}}\rho\big|_{(\mathrm{H_2+CO})_{\mathrm{max}}}}\frac{1}{s_{\mathrm{L}}} \qquad \text{eq. 2.43e}$$

$$\delta_{\mathrm{L},HRR,\mathrm{max}} = \frac{\lambda_{\mathrm{th}}\big|_{HRR_{\mathrm{max}}}}{c_{\mathrm{p}}\big|_{HRR_{\mathrm{max}}}\rho\big|_{HRR_{\mathrm{max}}}}\frac{1}{s_{\mathrm{L}}} \qquad \text{eq. 2.43f}$$

$$\delta_{\mathrm{L,{}^\bullet OH,max}} = \frac{\lambda_{\mathrm{th}}\big|_{\mathrm{{}^\bullet OH_{max}}}}{c_{\mathrm{p}}\big|_{\mathrm{{}^\bullet OH_{max}}}\rho\big|_{\mathrm{{}^\bullet OH_{max}}}}\frac{1}{s_{\mathrm{L}}} \qquad \text{eq. 2.43g}$$

$$\delta_{\mathrm{L,therm}} = z(T_{\mathrm{max}}\cdot 0.95) - z(T_{\mathrm{min}}\cdot 1.05) \qquad \text{eq. 2.43h}$$

2.2.6 Turbulent Flames

For SI engine combustion, no laminar flame front will form. The presence of turbulence and charge motion will perturb the flame front and increase its

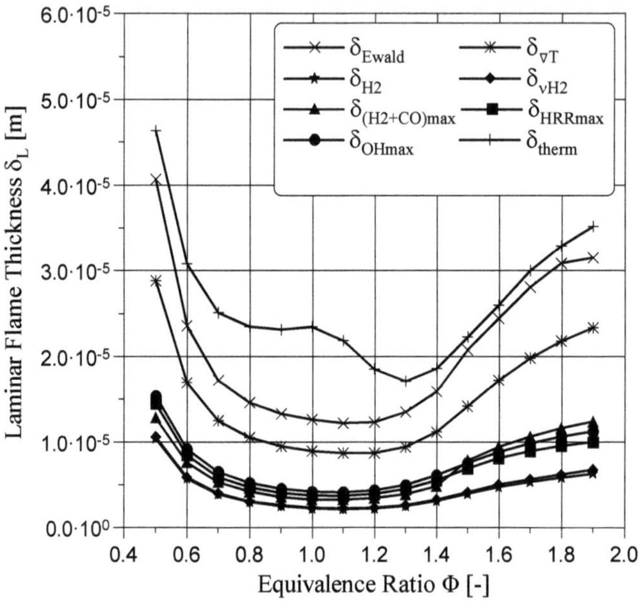

Figure 2.5: Comparison of laminar flame thicknesses calculated by multiple definitions. Variation of the equivalence ratio ϕ for the combustion of Cyclopentanone at 800 K and 50 bar.

surface area. Heat and mass transport rates and the effective burning velocity are accelerated. To build models for turbulent flame propagation, turbulent flame speeds have to be determined experimentally and then correlated to known or observed influence parameters (see 2.3.1).

Turbulent Flame Speed Measurement

Turbulent flame speeds are measured similarly to laminar burning velocities (see 2.2.5). The preconditioned mixture is put into a spherical combustion vessel with optical access. Then, an isotropic turbulence field with a controlled turbulence u' is created in the combustion chamber using fans. The mixture is then ignited with a defined ignition energy and the flame propagation is tracked by an optical setup using the *Mie* scattering method. Additionally, *Schlieren*-Imaging can be used. In post-processing, the proportions of unburnt masses m_{ub} and burnt masses m_{b} are taken by applying the method schematically drawn in figure 2.6: The single layer is divided into three radii; the root radius r_r, whose corresponding circle contains only burnt mass; the tip radius r_t, which contains all burnt mass, and a reference radius r_f. While the first two are clearly expressed, multiple definitions exist for the latter [11]. By integrating the flame front curve within the three integration limits (radii), the mass proportions, their conversion rates, and ultimately the turbulent flame speed can be calculated. The whole principle, the assumptions taken and its limitations are discussed in more detail by *Bradley et. al* [11].

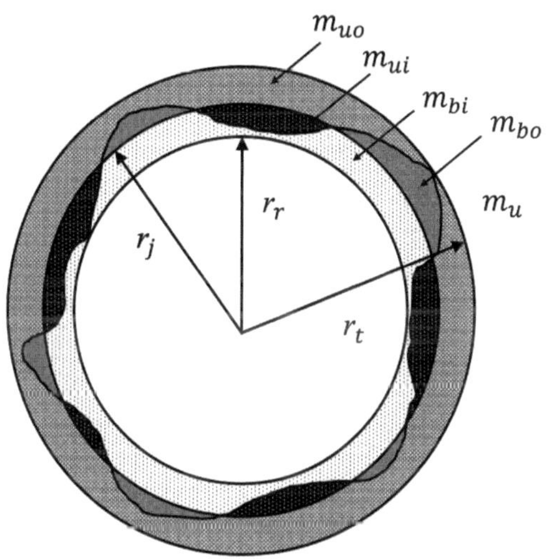

Figure 2.6: Wrinkling of the flame front in a turbulent flame. Recreated from [11].

Hydrodynamic Instability

Darrieus [24] and *Landau* [70] both independently recognized the existence of hydrodynamic instabilities in the flame fronts of premixed flames. According to them, in a perturbed flame, the thermal gas expansion caused by the exothermal reactions in the flame increases the normal component of the flames' laminar propagation speed, with the laminar burning velocity being constant. This leads to hydrodynamic disturbances by local outgrowths that increase the flame surface of large-scale laminar flames. This effect is further described by *Law* in [71]: It is assumed that the streamtube diameters are the same size upstream and downstream of the flame, since there is no disturbance present. The thermal expansion leads to a normal downstream flow velocity higher than that of the upstream, with the tangential components staying the same (see arrow diagram in the upper part of figure 2.7). The streamtube of diameter A is widened in the vicinity of the flame surface, causing it to slow down. Since the laminar flame speed is unchanged, the difference in the velocities of the approaching mass and the flame causes the flame front to bend inwards into the unburnt regime. The opposite effects show in the burnt area. As those effects are self-sustaining as well as self-amplifying, this phenomenon is unquestionably unstable. Referring to the first observers of this phenomenon, this effect is called *Darrieus-Landau Instability*.

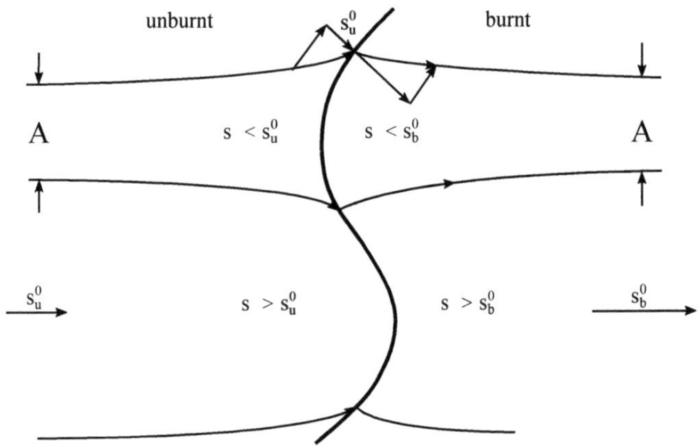

Figure 2.7: Hydrodynamic instabilities in a laminar flame. Recreated from [71].

Diffusional–Thermal Instability

Especially for the combustion of fuel(-mixtures) with a thermal diffusivity greater than the mass diffusivity (*Lewis*-Numbers > 1), diffusional-thermal instabilities show for perturbed flame fronts [71]. As schematically depicted in figure 2.8, on the convex segment (upper part of the flame), the heat diffusion into the unburnt mixture is exaggerated while the incoming mass is spread over a wider flame area, reducing the effective mass flux into the flame. For the concave segment, reverse effects apply. This ultimately leads to the stabilization and un-wrinkling of the flame. For *Lewis*-Numbers smaller than 1, the opposite behavior occurs and the flame mode becomes unstable. To allow for a prediction of the behavior of different fuels, their Lewis numbers have to be calculated for combustion engine relevant boundary conditions (see section 3.2.3).

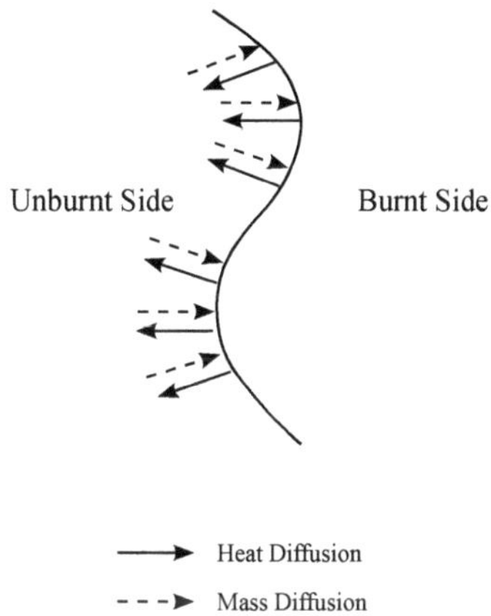

Figure 2.8: Diffusional-thermal instabilities in a laminar flame. Recreated from [71].

Lewis and Markstein Number

To characterize the relationship between thermal diffusivity and mass diffusivity, the *Lewis*-Number *Le* is introduced. It is calculated using the eq. 2.44 [81]. The *Markstein*-Number additionally takes the flame stretch into account and is calculated by dividing the *Markstein*-Length $\mathcal{L}_b$ by the laminar flame thickness δ_L. This correlation is based on the findings of *Markstein* for the stability of propagating flames [79][80].

$$Le = \frac{\text{thermal diffusivity}}{\text{mass diffusivity}} = \frac{\lambda_\mathrm{th}/c_\mathrm{p}}{\rho D_\mathrm{F}} \qquad \text{eq. 2.44}$$

$$Ma = \frac{\mathcal{L}_b}{\delta_\mathrm{L}} \qquad \text{eq. 2.45}$$

The publications of *Chaudhuri et al.* [17], *Han et al.* [47], *Hann* [48], and *Nguyen et al.* [86] show, that *Lewis*- and *Markstein*-Numbers are almost interchangeable. As the acquisition of local stretch rates of arbitrary mixtures cannot directly be done by simulating a freely propagating flame, *Le* has been chosen to compare the sensitivity of flame wrinkling for changing fuel(-mixtures) (see also, [48]). Although *Hann*'s experience seems to indicate the opposite (also, *Yang et al.* [118] and *Sivashinsky et al.* [106]), it has to be considered, that the effects on wrinkled flames are also influencing the propagation speed of turbulent flames. It has yet to be checked, if the effects observed for methane and gasoline combustion are altered by different fuel characteristics. Bigger differences in their *Lewis*-Numbers might extend to variations of diffusive phenomena in the flame.

2.2.7 Ignition Delay Measurements

Laboratory ignition delay measurements are usually divided into two groups depending on the temperatures at which they are gathered. For high temperature regimes, Shock Tube (ST)-measurements are used. For lower temperatures, Rapid Compression Machines (RCMs) are utilized.

STs consist of a long, cylindrical vessel divided into two sections by a diaphragm: the driving and the driven section. While the driven section is filled with the mixture to be tested, the driver section is pressurized with a driver gas until the diaphragm bursts when reaching the desired pressure. The expanding driver gas then compresses the test gas to a base pressure. The time between reaching the operation point and a recognizable pressure increase is the measured ignition delay [25]. Due to their design, shock tubes are not able to detect long ignition delay times and thus are typically bound to high temperature and high pressure operating points ($\sim 10-50$ bar, up to 1400 K).

Lower temperatures, on the other hand, are covered by RCM measurements. They consist of three sections: a pneumatic tank, a hydraulic transmission, and the testing reactor itself. With pneumatic pressure and the hydraulic transmission, the test gas is compressed up to a fixed point where the compression cylinder is locked in position. Again, the time until the chamber's pressure rises is measured and taken as the ignition delay time [109]. Due to their construction, RCMs come close to the ideal of an adiabatic compression stroke. They are usually used for temperatures up to 900 K [40] [100].

An example of the pressure rise in an RCM is given in figure 2.9. The initial pressure rise caused by the compression stroke of the cylinder, the slight decay in pressure due to energy loss through the vessel's walls, and the high gradient pressure rise by auto-ignitions are clearly visible. The calculation of Ignition Delay Times (IDTs) can be done in model reactors with reaction kinetic calculation. Further information can be found in chapter 3.

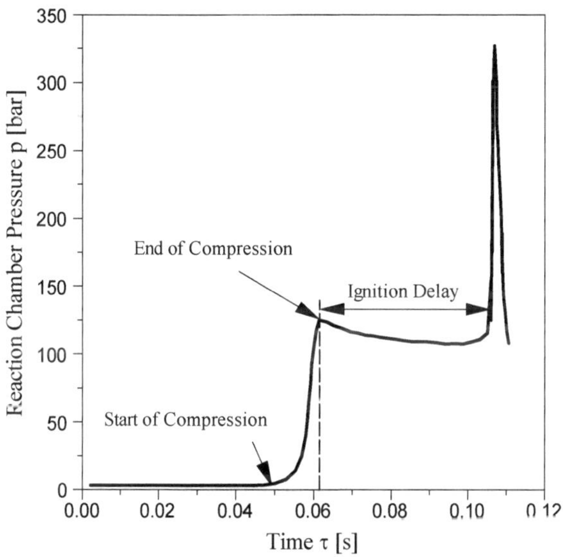

Figure 2.9: Interpretation of an RCM pressure trace for a CH_4/air mixture of $\phi = 0.526$, $p = 125.8$ bar, $T = 887$ K. Data taken from [100].

2.2.8 Flammability Limits

The boundary conditions at which combustion cannot be kept up (or started at all) are called flammability limits. They exist for the rich (e.g. for methane $\phi > 1.5$) and lean (methane $\phi < 0.5$) mixtures [92] [113]. Any mixture outside those limits will not build a stable, self-sustaining flame and in consequence will not burn [58]. The limits are called rich (*upper*) or lean (*lower*) flammability limits, accordingly. They are measured in standardized tubes or in closed cylindrical vessels [102].

The flammability limits of the following table have been taken from [102] and translated into Equivalence Ratios (ERs) and AFRs. For mixtures such as the Toluene (Primary) Reference Fuel (TRF) used in this work, the lower flammability has been determined by the Law of *Le Châtelier*, following eq. 2.46. It states, that the inverse of the flammability limit of a given mixture FL_{mix}, is equal to the sum of the ratio between mole fraction x_{i} and the flammability limit FL_{i} of each mixture component [16]. In their work, *Mendiburu* et al. showed that this procedure is reasonable for the lower flammability limits but yields unrealistically high values for the Upper Flammability Limit (UFL). As for this work, the lean burn limits of the used fuels are of greater interest, the UFLs have not been calculated for mixtures.

$$LFL_{\mathrm{mix}} = \frac{1}{\sum_{i=1}^{N} \frac{x_{\mathrm{i}}}{LFL_{\mathrm{i}}}} \qquad \text{eq. 2.46}$$

2.3 Fundamentals of Engine Combustion Modeling

Engine combustion simulation is defined as the calculation of combustive (or linked) processes by using empiric or phenomenological models. For this, multiple approaches with different dimensional orders can be used. For detailed, Three-Dimensional (3D) modeling, the combustion chamber and all peripherals are divided into equal sub-volumes, solving fundamental continuity equations for each volume and calculating mass- and energy flux at the borders between them. As this is connected to high computational and monetary efforts, especially for modelling real driving behaviour and other transient scenarios, models of lower dimensions are often preferred. In a 1D simulation, every volume is replaced by a pipe of similar (or, if mirroring a pipe, exact) diameter and length (and thus: volume). Solving the initial problem is then reduced to local discretization and the boundaries between volumes. For those simulations, the combustion is often scaled down to a 0D or QD problem, where either no (0D) or reduced (QD) local resolution exists. For QD calculations, all combustive processes are described in dependence on the flame radius r_{f}.

Table 2.2: Flammability limits for the used fuels (taken from [102]) and fuel mixtures (calculated).

Fuel (-mixture)		LFL			UFL	
	vol-%	ER	AFR	vol-%	ER	AFR
Isooctane	1.00	0.603	1.659	7.00	4.491	0.223
Toluene	1.10	0.478	2.093	7.10	3.283	0.305
n-Heptane	1.10	0.584	1.712	7.00	3.952	0.253
Ethanol	3.30	0.527	1.897	20.0	2.616	0.382
CPN	1.30	0.427	2.344	10.8	3.139	0.319
AN	1.20	0.507	1.972	6.30	2.489	0.402
DMC	3.80	0.661	1.512	21.3	1.979	0.505
MeFo	4.50	0.551	1.815	23.0	1.763	0.567
Hydrogen	4.00	0.099	10.06	75.0	7.160	0.140
TRF	1.04	0.562	1.779	-	-	-
TRFE5	1.13	0.561	1.782	-	-	-
TRFE10	1.22	0.559	1.788	-	-	-
E10CPN50m	1.26	0.491	2.038	-	-	-
E10AN50m	1.21	0.534	1.871	-	-	-
E10CPN50v	1.27	0.484	2.064	-	-	-
E10AN50v	1.21	0.531	1.883	-	-	-
E5CPN	1.15	0.548	1.823	-	-	-
E5AN	1.14	0.556	1.800	-	-	-
E10CPN	1.23	0.550	1.818	-	-	-
E10AN	1.22	0.556	1.799	-	-	-
DMC65MeFo35	4.07	0.620	1.612	-	-	-

For engine modeling, multiple quantities, variable both in time and location, have to be known. Mass and energy transport can occur across the system boundaries of the combustion chamber. As shown in figure 2.10, the incoming and outgoing masses dm_{in}, dm_{ex}, and dm_{bb} have to be known. The energy flux is divided into the energy provided by the combustion of the fuel dQ_{F}, the heat losses through the cylinder walls, piston deck and cylinder roof dQ_W, and the work delivered to the cranktrain dW. All relations between those quantities are defined by the laws of continuity and thermodynamic equation of state (see 2.1).

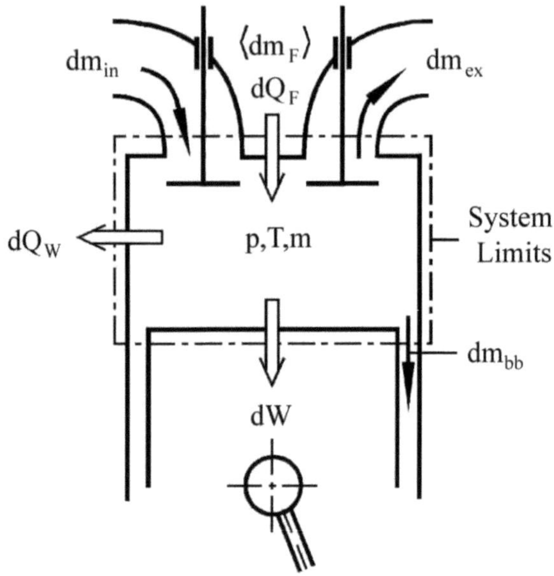

Figure 2.10: Schematic overview of the thermodynamic system *combustion chamber*. Mass and heat fluxes across the system border. Translated from [98].

One major input value for the combustion simulation is the cylinder mass and its composition (stoichiometric ratio, Exhaust Gas Recirculation (EGR)-ratio, Water Rate (WR)). In the case of 0D and QD simulation, their quantities are usually acquired by simulating the load exchange through the valves by means of 1D models. They are based on the fundamental equations for instationary, one-dimensional, compressible string flows, including wall heat and wall friction submodels. If the in- and outgoing flows and their turbulence are known, the cylinder load at the moment the inlet valve(s) close can be calculated. In this work, all 1D simulation is either done with *Gamma Technologies GT-Power*-software or with the *FKFS* cylinder module [43].

The most basic goal of engine simulation is the calculation of the engine torque at a specific operating point. Often, the equivalent of the Indicated Mean Efficient Pressure (IMEP) is used, as it allows for a better comparison of engines of different displacements. The IMEP is the area within the thermodynamic cycle in a p/V-diagram, and thus cylinder pressure and volume need to be known for its calculation. The cylinder volume is obtained by bore and current stroke (by crank drive kinematics). The cylinder pressure is the result of overlaying the motored cylinder pressure with the overall net energy put into the system. Ultimately, this is based on a modeling of the burn rate dQ_b. It is defined as the conversion of chemical fuel energy into heat with time and can be described either by post-processed measurement data (Pressure Trace Analysis (PTA)), empirical (e.g. *Vibe*) or phenomenological models. Empirical models describe the combustion process by using substitute equations. They can mirror the burn rate when calibrated correctly, but can only be used within a narrow operation range around their calibration point. Phenomenological models, on the other hand, try to mimic all relevant physiochemical fundamentals as closely as possible while still being performant. Apart from the entrainment model used in this work (see 2.3.1) [4], a fractal [10] and a flame density model [101] can be used. An overview of the more common phenomenological burn rate models is given by *Demesoukas et al.* [26]. As *Hann* formulated, all phenomenological combustion models rely on the same physical inputs, making the process proposed in chapter 4 viable for all of them [48].

2.3.1 Entrainment Model

As with most phenomenological models, the entrainment model assumes the formation of a hemispherical flame propagating from the spark plug. The cylinder itself is segregated into two zones, one containing all the burnt mass m_{b}, the second all the unburnt mass m_{ub} (see schematic in figure 2.11). The flame front, which is not modeled as a discrete thermodynamic zone, separates them.

To get the enthalpy flux from combustion (eq. 2.47), the conversion rate into burnt masses $\frac{dm_{\mathrm{b}}}{dt}$ and the lower heating value of the fuel LHV are needed. While the lower heating value is an intrinsic fuel quantity (see subsection 2.1.4), the mass conversion rate is modeled according to eq. 2.48: The mass flow into the burnt zone equals the conversion rate of the mass in the flame zone m_{f} during the characteristic burn up time τ_{L}. This can be rewritten to

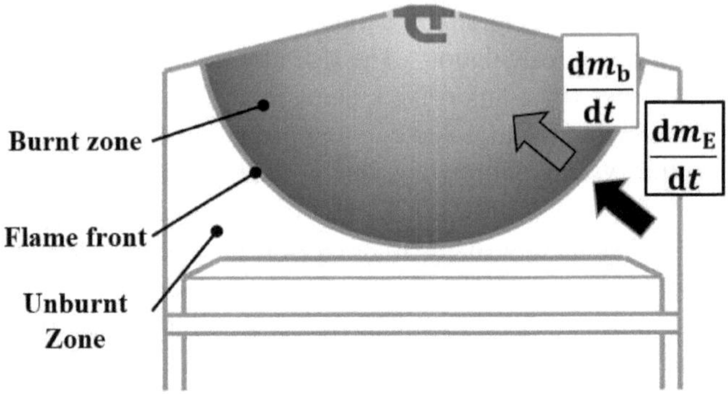

Figure 2.11: Definition of the two thermodynamic zones in the entrainment model.

the second term in eq. 2.48, since the mass in the flame is dependent only on the entraining mass flow $\frac{\mathrm{d}m_\mathrm{E}}{\mathrm{d}t}$ and the mass flow into the burnt zone.

$$\mathrm{d}Q_\mathrm{b} = \frac{\mathrm{d}m_\mathrm{b}}{\mathrm{d}t} LHV \qquad \text{eq. 2.47}$$

$$\frac{\mathrm{d}m_\mathrm{b}}{\mathrm{d}t} = \frac{m_\mathrm{F}}{\tau_\mathrm{L}} = \frac{m_\mathrm{E} - m_\mathrm{b}}{\tau_\mathrm{L}} \qquad \text{eq. 2.48}$$

The characteristic burn up time is defined by the *Taylor* microscale l_T over the laminar flame speed s_L following eq. 2.49. The *Taylor* microscale is calculated by eq. 2.50 by combining the *Taylor* factor X_Taylor (calibration constant) with the turbulent kinetic viscosity ν_T, the integral length scale l_int, and the turbulent fluctuation velocity u'.

$$\tau_\mathrm{L} = \frac{l_\mathrm{T}}{s_\mathrm{L}} \qquad \text{eq. 2.49}$$

$$l_\mathrm{T} = \sqrt{X_\mathrm{Taylor} \frac{\nu_\mathrm{T} \cdot l_\mathrm{int}}{u'}} \qquad \text{eq. 2.50}$$

The entraining mass flow into the flame zone $\frac{dm_E}{dt}$ is then given by eq. 2.51. It is formulated as the product of the density of the unburnt mixture ρ_{ub}, the flame surface area A_F, and the turbulent flame speed s_T. The latter is calculated depending on the used model, where a large variance exists ([23] [77] [85] [95] [121]). *Hann* has published a comprehensive comparison between them [48]. As only two turbulent burn rate models - the ones of *Damköhler* [23] and *Peters* [96] - have been used, those are explicitly explained now. Additional information on burn rate models and an extensive comparison study can also be found in [76].

$$\frac{dm_E}{dt} = \rho_{ub} \cdot A_F \cdot s_T \qquad \text{eq. 2.51}$$

Damköhler Model

The *Damköhler* model uses a simple approximation for the turbulent flame speed by adding up the turbulent fluctuation velocity u' and the laminar flame speed s_L (eq. 2.52)[23]. An extended version can be used by adding an exponent to it (eq. 2.53). This serves two purposes: For exponents greater than 1, the unrealistic behavior of turbulent flame propagation directly proportional to the laminar flame speed is corrected and it adds a calibration factor to the model.

$$s_{T,\text{Damköhler}} = u' + s_L \qquad \text{eq. 2.52}$$

$$s_{T,\text{Damköhler,e}} = (u' + s_L)^n \qquad \text{eq. 2.53}$$

Peters Model

The model proposed by *Peters* [95] is built off of eq. 2.54. Again, the laminar burning velocity s_L is used. Additionally, the integral length scale l_{int}, the laminar flame thickness δ_L (see 2.2.5), and the *Damköhler*-number *Da* influence the turbulent combustion.

$$s_{\mathrm{T,Peters}} = s_{\mathrm{L}}\left(1 + 0.195\frac{l_{\mathrm{int}}}{\delta_{\mathrm{L}}}\left(\sqrt{1 + \frac{20.5}{Da}} - 1\right)\right) \quad \text{eq. 2.54}$$

The *Damköhler*-number Da is defined by the laminar flame speed s_{L}, the turbulent fluctuation velocity u', the integral length scale l_{int}, and the laminar flame thickness δ_{L} (eq. 2.55)

$$Da = \frac{s_{\mathrm{L}}}{u'}\frac{l_{\mathrm{int}}}{\delta_{\mathrm{L}}} \quad \text{eq. 2.55}$$

2.3.2 Knock Models

A major limitation on the path to more efficient ICEs is the occurrence of knocking combustion. To better predict unwanted engine behavior in the design phase of an engine, auto-ignition and engine knock models have to be applied. Since the two-stage auto-ignition model presented by *Fandakov et al.* [36] has been extended for this work, the original version is introduced shortly. Furthermore, knock models are used to check if with an auto-ignition also knocking combustion is present. The two utilised knock models, the one of *Fandakov* [34] and the one of *Hess* [53] are briefly explained. They have both been created for gasoline fuels or gasoline/ethanol mixtures. Their applicability for other fuels and mixtures has not been clarified yet.

Two-Stage Auto-Ignition Model

The most commonly used model for the prediction of auto-ignition occurrence in ICEs was originally formulated by *Livengood* and *Wu* [78]. They used an integral to sum up inverse IDT data (eq. 2.56) beginning at Inlet Valve Closes (IVC) (or, depending on the source, 90 °CA before Top Dead Center (TDC)) and ending with End of Combustion (EOC). If the IDTs are modelled correctly (see 2.3.5), this integral can be easily calculated during simulation runtime by adding to it in every iteration. Should the integral reach a value of 1 before the EOC is reached, the simulated operating point will show auto-ignition behavior.

$$1 = \int_{t_{IVC}}^{t_{EOC}} \frac{1}{\tau} \mathrm{d}t \qquad \text{eq. 2.56}$$

Pan et al. first proposed the use of two sequential *Livengood-Wu*-integrals to properly describe the behavior of fuels with a significant two-stage ignition [90]. While they only applied it to gasoline Homogeneous Charge Compression Ignition (gHCCI) combustion simulation, *Fandakov et al.* repurposed this idea for standard gasoline SI combustion, including the influence of changing AFR and EGR rates [36]. With this, eq. 2.56 is changed to eq. 2.57 for the low temperature auto-ignition and a second, high-temperature ignition integral is introduced (eq. 2.58). The acronym for the Boundary Conditions (BCs) in the equations refers to changing conditions in terms of temperature, pressure, AFR, EGR rate, or WR.

$$1 = \int_{t_{IVC}}^{t_1} \frac{1}{\tau_{\mathrm{low}}(BC)} \mathrm{d}t \qquad \text{eq. 2.57}$$

$$1 = \int_{t_1}^{t_{EOC}} \frac{1}{\tau_{\mathrm{high}}(BC, T_{\mathrm{incr}}(BC), p_{\mathrm{incr}}(T_{\mathrm{incr}}))} \mathrm{d}t \qquad \text{eq. 2.58}$$

The three input parameters, τ_{low}, τ_{high}, and T_{incr} are gathered from reaction kinetics simulations beforehand and taken from correlation equations during the simulation runtime. Their acquisition is further described in section 3.3, the correlations are shown in subsection 2.3.5. The pressure increase by the first stage of the ignition is calculated by using the thermodynamic equation of state and the temperature increase by the first stage.

Fandakov Knock Model

To achieve better simulation results for the knock onset prediction of gasoline and gasoline/ethanol combustion, especially with added EGR at high engine loads, *Fandakov* published a new knock modeling approach [35]. His idea was to model the mass fraction of unburnt gas in the cool thermal boundary layer with the assumption that the higher the mass fraction in the cool layer is, the lower the probability of knock occurring will be. This approach was introduced since popular knock occurrence criteria wrongfully assumed that no knock could occur after a certain fraction of mass had been burnt.

The unburnt mass fraction in the boundary layer $w_{\text{ub,bl}}$ is given by eq. 2.59. The mass averaged temperature of the boundary layer T_{bl} is assumed to be the mean value of the temperature in the unburnt zone T_{ub} and the cylinder wall temperatures. The volume of the unburnt mass in the boundary layer $V_{\text{ub,bl}}$ is estimated by discretizing the combustion chamber into smaller volumes and taking the thermal boundary layer thickness δ_{t} into account. To calibrate the model, the threshold value for $w_{\text{ub,bl}}$ is set to a value where measured Knock Limitation Spark Advances (KLSAs) are met. The whole model is described in detail in [35].

$$w_{\text{ub,bl}} = \left(\frac{T_{\text{bl}}}{T_{\text{ub}}} \left(\frac{1}{V_{\text{ub,bl}}} - 1 \right) + 1 \right)^{-1} \qquad \text{eq. 2.59}$$

Hess Knock Model

When having problems with reaching desired accuracies for predicted KLSAs, *Hess et al.* developed an additional knock occurrence criterion [53][54] for use with gasoline combustion. They found, that with auto-ignition being present for operating points with late ϕ_{50}, the state of the pre-reactions taking place in the unburnt gas has further progressed what needs to be accounted for when calculating knock onset. *Hess* is basing one of his newly developed knock occurrence criteria on the work of *Kleinschmidt*, who did dimensional analysis on pressure oscillations caused by auto-ignition [66]. This leads to the formulation of the dimensionless "auto-ignition" Mach number Π_{AI}. Its definition is shown in eq. 2.60.

$$\Pi_{\text{AI}} = \frac{A \rho_1 e^{(-\beta_\pi / T_1)} V_1^{1/3}}{M_{\text{fuel}} a_{\text{s}1}} \qquad \text{eq. 2.60}$$

The parameters A, β_π, and M are fuel related and dropped by *Hess* for his criterion. Additionally he measured the influence of the remaining expression $e^{(1/T_1)}$ as neglectable for higher temperatures. Adding the state of the mixture leads to eq. 2.61. It is solved at the point of auto-ignition $\Pi_{\text{AI},0}$ and compared to its state at the start of the high pressure simulation $\Pi_{\text{AI,start}}$ (eq. 2.62). The ratio between them can be interpreted as an artificial measure of the pressure amplitude caused by the auto-ignition. If it is beyond a calibrated threshold value Π_{KB}, the operating point is regarded as knock limited. As this criterion was validated with four different qualities of gasoline (RON98, RON95E10, RON95E20, and RON95M20), it shows potential for use with other additives and fuels.

$$\Pi_{AI,0} = p_{ub} \frac{\rho_{ub} V_{ub}^{0.5}}{\sqrt{\kappa_{ub} R_{ub} T_{ub}}} \qquad \text{eq. 2.61}$$

$$\Pi = \frac{\Pi_{AI,0}}{\Pi_{AI,start}} \qquad \text{eq. 2.62}$$

2.3.3 Laminar Burning Velocity Correlations

To generate laminar burning velocity data within conditions where no measurements are possible, both *Gülder* [45] and *Heywood* [55] developed their correlations based on eq. 2.63. They were intended for use with gasoline and the correlation parameters given in eq. 2.64 (*Gülder*) and eq. 2.65 (*Heywood*). They calibrated them with measurements at low pressures and temperatures and extrapolated the data from there.

$$s_L(\phi,T,p,F) = s_{L,0}(\phi)\left(\frac{T_{ub}}{T_0}\right)^{\alpha}\left(\frac{p}{p_0}\right)^{\beta}(1 - F \cdot w_{EGR}^{n}) \qquad \text{eq. 2.63}$$

$$s_{L,0}(\phi) = 0.422 \cdot \phi^{0.15} \cdot e^{-5.18 \cdot (\phi - 1.075)^2} \qquad \text{eq. 2.64}$$

$$\alpha = 2$$
$$\beta = -0.5$$
$$F = 2.5$$
$$n = 1$$

$$s_{L,0}(\phi) = 0.305 - 0.549 \cdot (\phi - 1.21)^2 \qquad \text{eq. 2.65}$$

$$\begin{aligned}\alpha &= 2.18 - 0.8 \cdot (\phi - 1)\\ \beta &= -0.16 + 0.22 \cdot (\phi - 1)\\ F &= 2.06\\ n &= 0.77\end{aligned}$$

Müller et al. [84] published a method to approximate the laminar burning velocities of hydrocarbon fuels at over-stoichiometric conditions with algebraic functions. Its base correlation is limited to the narrow range of boundary conditions given in 2.3. Later, this model was extended by *Ewald* [33] to better predict the behavior for changing AFRs by adding spline functions to replace the existing polynomials. *Hann* added functionality for the inclusion of binary mixtures [49], and the author later extended it again to be able to mirror the influences of externally added water [22].

Table 2.3: Boundary limits of the *Müller* base correlation [84].

Value	from	to	unit
T_{ub}	298	800	K
p	1	40	bar
λ	1	1.7	
w_{EGR}	0	0	
w_{H2O}	0	0	

Equation eq. 2.66 describes the fundament of the *Müller* correlation. The frequency factor of the reaction $A_{T,0}$ is defined in eq. 2.67, the reactive mass fraction w_r in eq. 2.68. Additionally, the temperatures of the unburnt zone T_{ub}, the reaction zone T_0, and the burnt zone T_b are taken into account.

$$s_L = A_{T,0} \cdot w_r^m \cdot \left(\frac{T_{ub}}{T_b}\right)^r \cdot \left(\frac{T_b - T_0}{T_b - T_{ub}}\right)^n \qquad \text{eq. 2.66}$$

The frequency factor $A_{T,0}$ is dependent on the two calibration parameters F and G as well as on the temperature in the reaction zone T_0.

$$A_{T,0} = F \cdot e^{-\frac{G}{T_0}} \qquad \text{eq. 2.67}$$

The reactive mass fraction is the relationship between the reactive masses and the overall mass. It is dependent on the mixture fraction Z^*, the stoichiometric mixture fraction $Z^*{}_{st}$, and the EGR mass fraction w_{EGR}. A calibration of the AFR and EGR influence is done by the two parameters, n and n_{EGR}. The mixture fraction Z^* is explained by eq. 2.69, for stoichiometric conditions it is referred to as $Z^*{}_{st}$.

$$w_r(Z^*, w_{EGR}) = \left(\frac{Z^*}{1.1 \cdot Z^*{}_{st}}\right)^{n_a} \cdot \left(\frac{1 - Z^*}{1 - 1.1 \cdot Z^*{}_{st}}\right)^{\left(\frac{1}{1.1 \cdot Z^*{}_{st}} - 1\right) \cdot n_a} \cdot (1 - w_{EGR})^{n_{EGR}} \quad \text{eq. 2.68}$$

$$Z^* = \frac{m_{fuel}}{m_{fuel} + m_{air}} \quad \text{eq. 2.69}$$

Equation eq. 2.70 is used to model the temperature of the reaction zone T_0. As mentioned with the calculation of the laminar flame thickness, in reaction kinetics, multiple definitions for this zone exist. Over the course of the current work, the point of maximum heat release will be used (see 3.2.2). The parameters E_i and B_i allow for an adjustment of the pressure dependence. The two splines $S_1(Z^*)$ and $S_2(Z^*)$ are formulated in dependency of Z^* and have to be adjusted to fit the cross correlation between pressure and AFR.

$$T_0 = T_{ub} \cdot S_1(Z^*) + \frac{E_i \cdot S_2(Z^*)}{ln(\frac{B_i}{p})} \quad \text{eq. 2.70}$$

The same principle applies to the splines $S_3(Z^*)$ and $S_4(Z^*)$ in eq. 2.71. They are used to model the cross correlation between temperature and AFR. The temperature of the burnt mixture is also influenced by T_0 and the EGR rate referenced by w_{EGR}. *Hann* also introduced an extra exponential factor c to reduce the influence of the residual or recirculated burnt gases. It is examined if this is needed for the new fuels implemented [48].

$$T_b = T_{ub} \cdot (S_4(Z^*) \cdot (1 - w_{EGR}) + w_{EGR}) + (1 - w_{EGR})^c \cdot S_3(Z^*) \quad \text{eq. 2.71}$$

In [22] the author of the present work presented a method to accommodate the correlation results for the presence of additional water. The adjusted laminar burning velocity $s_{\mathrm{L,H2O}}$ is given by eq. 2.72.

$$s_{\mathrm{L,H2O}} = s_{\mathrm{L}} \cdot e^{-\left(b_{\mathrm{H2O}} - \frac{T_{\mathrm{ub}}}{c_{\mathrm{H2O}}}\right) \cdot w_{\mathrm{H2O}}} \qquad \text{eq. 2.72}$$

2.3.4 Laminar Flame Thickness Correlation

The laminar flame thicknesses have to be calculated with reaction kinetics simulation (see section 3.2.2) for the same reason as with the LBVs - no measurement data can be acquired for engine relevant pressures and temperatures. Depending on the complexity of the reaction mechanism and the boundary conditions, this can take up to an hour of single-core computing time per case. As acquiring a new laminar flame thickness for each new process step in the combustion simulation is not feasible, like the LBVs they are calculated beforehand and approximated by a correlation. According to *Peters*, laminar flame thicknesses can be estimated by their proportional correlation to the ratio between kinematic viscosity ν and the laminar burning velocity s_{L} (eq. 2.73) [96].

$$\delta_{\mathrm{L}} = \alpha_{\delta_{\mathrm{L}}} \frac{\nu}{s_{\mathrm{L}}} \qquad \text{eq. 2.73}$$

As there are no considerable dependencies between the kinematic viscosity and varying pressures, dilutions, or mixture compositions, the model assumes a mean kinematic viscosity ν for all those cases. Only the sensitivity to changing temperatures is addressed. This is pictured in the resulting eq. 2.74.

$$\nu = \frac{\eta}{\rho_{\mathrm{ub}}} = \frac{3.85 \cdot 10^{-7} T_{\mathrm{ub}}^{0.6774}}{\rho_{\mathrm{ub}}} \qquad \text{eq. 2.74}$$

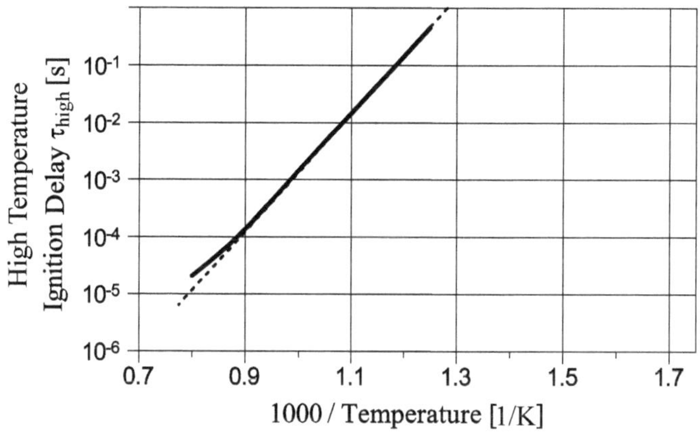

Figure 2.12: High temperature ignition delay of pure Anisole at 30 bar. The dashed line symbolizes a single Arrhenius fit.

2.3.5 Ignition Delay Correlation

Simple ignition delay correlations can be used for fuels that have an ignition delay behavior resulting in linear trend in the log(τ)/(1000 K / T) diagram. As an exponential expression perfectly mirrors this, Arrhenius-type equations like the one in eq. 2.35 are used for these cases. The temperature dependency is already covered by the base equation, and the pressure and mixture compensation is achieved by adjusting the prefactor A and the exponential factor B that resembles the activation energy (see section 2.2.2) in regards to external influences. This method is widely used and further described, for example, by *Sarathy et al.* [104].

$$\tau = Ae^{\left(\frac{B}{RT}\right)} \qquad \text{eq. 2.75}$$

Fandakov Ignition Delay Correlation

For more complex fuels, a 3-Arrhenius approach is used. An exemplary introduction to their application is given by *Blomberg et al.* [6]. The whole combustion engine's relevant temperature region is divided into three regimes with their corresponding ignition delay times. A high temperature

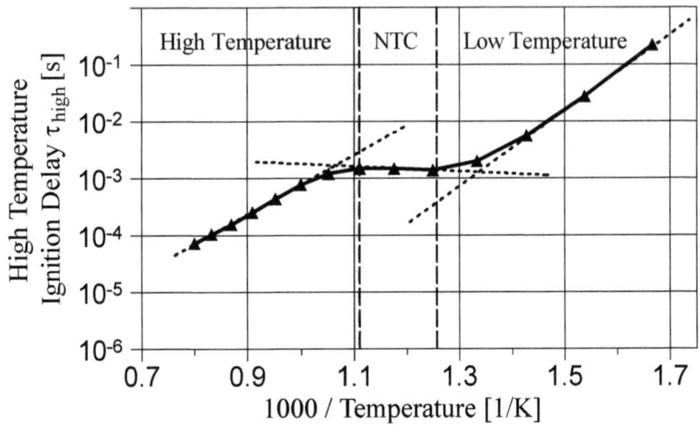

Figure 2.13: High temperature ignition delay of TRF at 30 bar. Division of the temperature dependency into three regimes; high temperature zone, low temperature zone and the NTC-zone.

zone (τ_3), a low temperature zone (τ_2) and one zone in between those two, the Negative Temperature Coefficient (NTC)-zone (τ_2). The three zones each get their own Arrhenius-type equation based on eq. 2.75 leading to eq. 2.80. The overall high temperature ignition delay then is given by eq. 2.79.

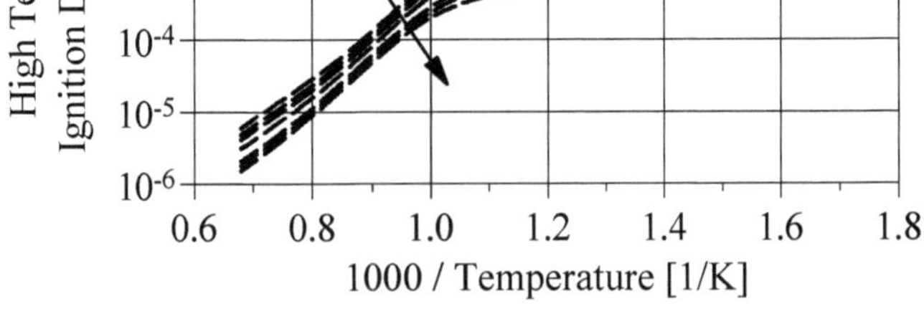

Figure 2.14: High-temperature ignition delay of $E10AN_{50m}$ for varying pressures.

$$\tau_{\mathrm{i,high}} = A_{\mathrm{i,high}} e^{\left(\frac{B_{\mathrm{i,high}}}{RT}\right)} \qquad \text{eq. 2.76}$$

With $A_{\mathrm{i,high}}$ and $B_{\mathrm{i,high}}$ both being dependent on the pressure in the unburnt mixture and the equivalence ratio as well as the EGR ratio as shown in eq. 2.77 and eq. 2.78. The pre factors are all dependent on the pressure and fitted to a second grade polynomial (S, P, Q, R), to two coupled exponential functions (U_{A}), or to a power function with offset (U_{B}).

$$\begin{aligned} A_{\mathrm{i,high}} =exp\Bigg(& U_{\mathrm{A,i}} + S_{\mathrm{lam,\,i}} \cdot (\lambda - 1)^2 + P_{\mathrm{lam}} \cdot (\lambda - 1)+ \\ & S_{\mathrm{EGR,i}} \cdot \left(\frac{m_{\mathrm{EGR}}}{m_{\mathrm{cyl}}} \cdot 100\right)^2 + P_{\mathrm{EGR,\,i}} \cdot \left(\frac{m_{\mathrm{EGR}}}{m_{\mathrm{cyl}}} \cdot 100\right)+ \\ & S_{\mathrm{tol,i}} \cdot \left(\frac{x_{\mathrm{tol}}}{x_{\mathrm{oct}} + x_{\mathrm{hep}}} - 0.514\right)^2 + P_{\mathrm{tol,i}} \cdot \left(\frac{x_{\mathrm{tol}}}{x_{\mathrm{oct}} + x_{\mathrm{hep}}} - 0.514\right)+ \\ & S_{\mathrm{oct,i}} \cdot \left(\frac{x_{\mathrm{oct}}}{x_{\mathrm{oct}} + x_{\mathrm{hep}}} - 0.223\right)^2 + P_{\mathrm{oct,i}} \cdot \left(\frac{x_{\mathrm{oct}}}{x_{\mathrm{oct}} + x_{\mathrm{hep}}} - 0.223\right)+ \\ & S_{\mathrm{etha,i}} \cdot \left(\frac{x_{\mathrm{etha}}}{x_{\mathrm{oct}} + x_{\mathrm{hep}} + x_{\mathrm{tol}}} - 0.115\right)^2 + \\ & P_{\mathrm{etha,i}} \cdot \left(\frac{x_{\mathrm{etha}}}{x_{\mathrm{oct}} + x_{\mathrm{hep}} + x_{\mathrm{tol}}} - 0.115\right)\Bigg)+ \end{aligned} \qquad \text{eq. 2.77}$$

$$\begin{aligned} B_{\mathrm{i,high}} =& U_{\mathrm{B,i}} + Q_{\mathrm{lam,\,i}} \cdot (\lambda - 1)^2 + R_{\mathrm{lam}} \cdot (\lambda - 1)+ \\ & Q_{\mathrm{EGR,i}} \cdot \left(\frac{m_{\mathrm{EGR}}}{m_{\mathrm{cyl}}} \cdot 100\right)^2 + R_{\mathrm{EGR,i}} \cdot \left(\frac{m_{\mathrm{EGR}}}{m_{\mathrm{cyl}}} \cdot 100\right)+ \\ & Q_{\mathrm{tol,i}} \cdot \left(\frac{x_{\mathrm{tol}}}{x_{\mathrm{oct}} + x_{\mathrm{hep}}} - 0.514\right)^2 + R_{\mathrm{tol,i}} \cdot \left(\frac{x_{\mathrm{tol}}}{x_{\mathrm{oct}} + x_{\mathrm{hep}}} - 0.514\right)+ \\ & Q_{\mathrm{oct,i}} \cdot \left(\frac{x_{\mathrm{oct}}}{x_{\mathrm{oct}} + x_{\mathrm{hep}}} - 0.223\right)^2 + R_{\mathrm{oct,i}} \cdot \left(\frac{x_{\mathrm{oct}}}{x_{\mathrm{oct}} + x_{\mathrm{hep}}} - 0.223\right)+ \\ & Q_{\mathrm{etha,i}} \cdot \left(\frac{x_{\mathrm{etha}}}{x_{\mathrm{oct}} + x_{\mathrm{hep}} + x_{\mathrm{tol}}} - 0.115\right)^2 + \end{aligned}$$

$$R_{etha,i} \cdot \left(\frac{x_{etha}}{x_{oct} + x_{hep} + x_{tol}} - 0.115 \right)$$

eq. 2.78

$$\frac{1}{\tau_{high}} = \frac{1}{\tau_{1,high} + \tau_{2,high}} + \frac{1}{\tau_{3,high}} \qquad \text{eq. 2.79}$$

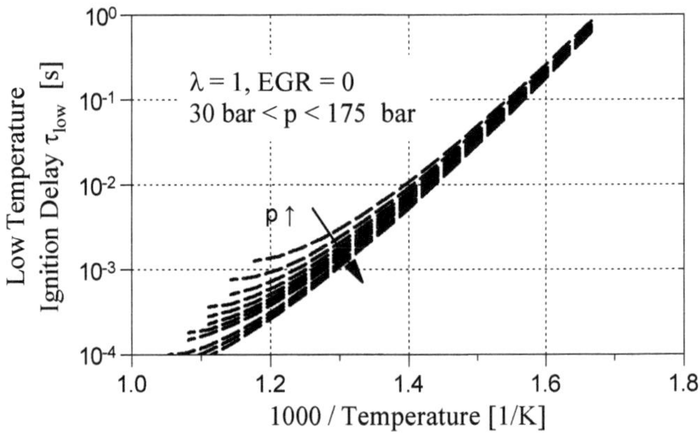

Figure 2.15: Low-temperature ignition delay of $E10AN_{50m}$ for varying pressures.

The same principle can equally be applied to the low temperature ignition delay τ_{low} if two-stage ignitions need to be addressed.

$$\tau_{i,low} = A_{i,low} e^{\left(\frac{B_{i,low}}{RT}\right)} \qquad \text{eq. 2.80}$$

$$\frac{1}{\tau_{low}} = \frac{1}{\tau_{1,low} + \tau_{2,low}} \qquad \text{eq. 2.81}$$

In addition to the known approaches, *Fandakov* extended the existing triple Arrhenius idea by modeling the temperature increase by the low temperature heat release T_{incr}, and adding a third correlation to the existing ones for low and high temperature ignition delays. To facilitate the fitting process, a helper variable $T_{incr,fit}$ has been added. Its relation to the low temperature T_{low} and the temperature increment is shown in eq. 2.82 [34]. The nomenclature is further visualized in figure 2.16.

$$T_{incr} = T_{incr,fit} \cdot 100 - T_{low} \qquad \text{eq. 2.82}$$

$$T_{incr,fit} = C_1\left(\frac{T_{low}}{100}\right)^4 + C_2\left(\frac{T_{low}}{100}\right)^3 + C_3\left(\frac{T_{low}}{100}\right)^2 + C_4\left(\frac{T_{low}}{100}\right) + C_5 \qquad \text{eq. 2.83}$$

with every coefficient C_i being determined by a function depending on pressure, mixture composition (surrogate composition, λ, EGR, and water content). The pressure dependence of the pre-factors S and P is again fitted to second grade polynomials, the one of U to two coupled exponential functions.

$$\begin{aligned} C_i =& U_{C,i} + S_{lam,i} \cdot (\lambda - 1)^2 + P_{lam,i} \cdot (\lambda - 1) + \\ & S_{EGR,i} \cdot \left(\frac{m_{EGR}}{m_{cyl}} \cdot 100\right)^2 + P_{EGR,i} \cdot \left(\frac{m_{EGR}}{m_{cyl}} \cdot 100\right) + \\ & S_{tol,i} \cdot \left(\frac{x_{tol}}{x_{oct} + x_{hep}} - 0.514\right)^2 + P_{tol,i} \cdot \left(\frac{x_{tol}}{x_{oct} + x_{hep}} - 0.514\right) + \\ & S_{oct,i} \cdot \left(\frac{x_{oct}}{x_{oct} + x_{hep}} - 0.223\right)^2 + P_{oct,i} \cdot \left(\frac{x_{oct}}{x_{oct} + x_{hep}} - 0.223\right) + \\ & S_{etha,i} \cdot \left(\frac{x_{etha}}{x_{oct} + x_{hep} + x_{tol}} - 0.115\right)^2 + \\ & P_{etha,i} \cdot \left(\frac{x_{etha}}{x_{oct} + x_{hep} + x_{tol}} - 0.115\right) \end{aligned}$$

eq. 2.84

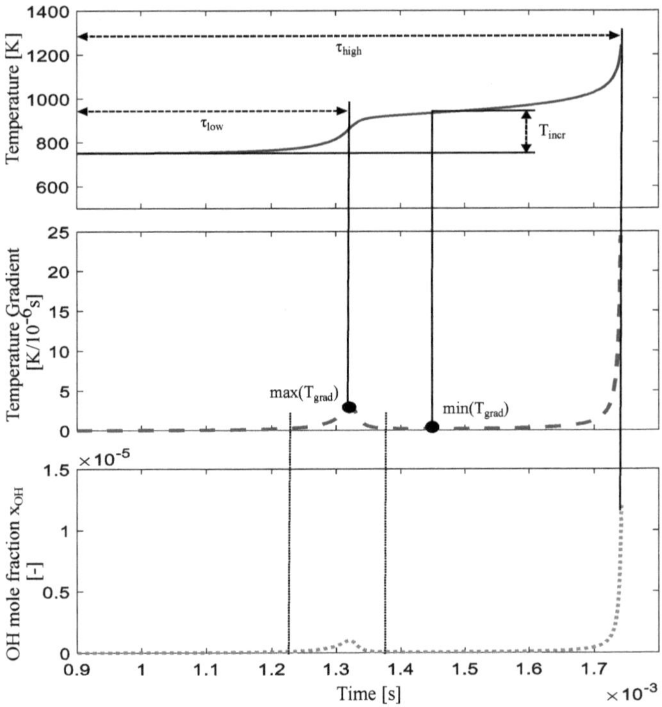

Figure 2.16: Definitions of variables of a two-stage ignition. Determination of τ_{low} and T_{incr} by the temperature gradient and τ_{high} by the OH radical concentration [34].

3 Reaction Kinetics Calculations

All reaction kinetic calculations are done with the open source software *Cantera* [41], developed by *Goodwin et al.*. It can be accessed through interfaces in *Matlab*, *Python*, and *Fortran*. The ease of use, together with its open source licensing, lead to the implementation of all calculation scripts in the Python programming language. This way, there are no costs associated with the presented methodology and every user has access to it.

The most fundamental task for all subsequent reaction kinetic calculations is rasterizing the boundary conditions and then determining the molecular (or mass-based) composition of the combustible mixture. As in the engine combustion simulation itself, for each calculation step, the cylinder pressure and temperature is known, the engine speed and load don't have to be considered. Also, since both the measurement and the turbulent flame speed calculation rely on the temperature of the unburnt mixture, other temperatures are only of minor interest. Thus, the remaining boundary conditions are: the temperature in the unburnt mixture T_{ub}, the average vessel pressure p, the mixture equivalence ratio ϕ, the mass-based EGR rate w_{EGR}, and the mass based water content w_{H2O}. Since the integration of external water injection seems to have already passed its phase of highest interest in the industry - even if it might see a renaissance for hydrogen combustion - the fitting of the influ ence of external water on reaction kinetic data is not done yet. A calculation as well as a fitting procedure for these datasets exist, but especially for the acquisition of laminar burning velocities for an additional dimension, the effort is too high for the small benefit expected.

3.1 Reaction Kinetic Mechanisms

In section 2.2, the importance to chose an appropriate mechanism for each fuel and fuel mixture has been highlighted. To give an orientation and allow for a complete reconstruction, the mechanisms used for all variants are in-

S. K. Crönert, *A Complete Methodology for the Predictive Simulation of Novel, Single- and Multi-Component Fuel Combustion*, Wissenschaftliche Reihe Fahrzeugtechnik Universität Stuttgart, https://doi.org/10.1007/978-3-658-43075-7_3

troduced. Additionally, the tables 3.1 and 3.2 show which mechanism has been used for which fuel or fuel mixture.

Table 3.1: Mechanism usage for LBV calculation depending on fuel composition. For fuels and fuel mixtures that were simulated with multiple mechanisms, the main mechanism is highlighted.

Fuel	Mechanism					
	Gri	Cai	UCSD	Dooley	CPN	AN
Hydrogen	X	**X**	X			
Methane	X	**X**	X			
Methanol	X	**X**	X			
Ethanol	X	**X**	X			
Methyl Formate				X		
Anisole						X
Cyclopentanone					X	
T(P)RF		**X**			X	X
+ Methanol		**X**			X	X
+ Ethanol		**X**			X	X
+ Cyclopentanone					X	
+ Anisole						X
+ Hydrogen		X				
DMC				X		
+ Methyl Formate				X		

As for the IDT calculations, mechanisms with a detailed Low Temperature Chemistry (LTC) are needed, and not all the mechanisms used for the LBV calculation are available in a version that supports this, so the matrix has to be adjusted for these uses. As both the new Cyclopentanone C_5H_8O (CPN) and Anisole C_7H_8O (AN) mechanisms are the most complex ones and support all other fuels used, they are chosen. Since those two mechanisms are quite similar, all simulations apart from those containing Anisole, Dimethyl Carbonate (DMC), or methyl formate are also done with the CPN mechanism.

Table 3.2: Mechanism usage for IDT calculation depending on fuel composition.

Fuel	Mechanism					
	Gri	Cai	UCSD	Dooley	CPN	AN
Hydrogen		**X**			X	
Methane		**X**			X	
Methanol		**X**			X	
Ethanol		**X**			X	
Methyl Formate		X				
Anisole						X
Cyclopentanone					X	
T(P)RF		**X**			X	
+ Methanol		**X**			X	
+ Ethanol		**X**			X	
+ Cyclopentanone					X	
+ Anisole						X
+ Hydrogen					X	
DMC		X				
+ Methyl Formate		X				

Gri 3.0 Mechanism

The Gri3.0 [107] is one of the most popular mechanisms and even ships with most of the reaction kinetics simulation software. It is designed and extensively validated for use with natural gases (mixtures of methane, ethane, and propane) and added proportions of hydrogen.

Cai Mechanism

In need of a mechanism that could appropriately model TRF surrogates, this one was created by *Cai et al.* [14]. It has been further extended since and is also able to simulate additional methanol or ethanol fuel content. Both the Anisole and the cyclopentanone mechanisms, and also the DMC and Methyl Formate (MF) mechanisms, have been created on top of the *Cai* mechanism base structure.

UCSD Mechanism

The mechanism from the University of California at San Diego was created for ethylene combustion in air [103]. Because it also includes other C_2 submechanisms, it can be used for hydrogen and methane as well as methanol and ethanol combustion simulations. It has to be mentioned that it was not designed for this use case and has not been calibrated to fit measurement data for those fuels.

Methyl Formate Mechanism

Pure Methyl formate is simulated with two mechanisms: The first was created by *Dooley et al.* [31] and has been further improved since [30]. Additionally, a submechanism for the simulation of DMC and MF has been engineered by *vom Lehn et al.* and included in the aforementioned *Cai*-mechanism [15] [73]. Since the latter includes low temperature chemistry, it is used for all calculations including methyl formate or dimethyl carbonate. The *Dooley*-mechanism is only used for validation and as a reference.

Anisole and Cyclopentanone mechanisms

In the course of the Forschungsvereinigung Verbrennungskraftmaschinen e. V. (FVV) project #1348 "Fuel Composition for CO2 Reduction" [12], two reaction mechanisms were created for Anisole and cyclepentanone combustion. This work has been done by the Institute for Technical Combustion (ITVTC) at North Rhine-Westphalia Technical University of Aachen (RWTH) Aachen, the ignition delay measurements were provided by Chair of High Pressure Gas Dynamics (HGD) (also RWTH). The mechanisms are based on the one created by *Cai et al.* [14] and include chemical reaction, thermodynamic, and transport data needed for the handling of Anisole or cyclopentanone combustion including intermediate species. This has been achieved by integrating the submechanisms for cyclopentanone [119] [120] and Anisole [13] into the base mechanism. The newly created reaction mechanisms will be referred to as *AN* or *CPN* as there is no other naming convention yet.

Species and Equation Counts

When reaction mechanisms have to include more complex species, the count of intermediate species also rises. The table in 3.3 shows, that with an increased species count, the equation count also rises non-linearly. As this directly resembles one dimension of the Differential Equation System (DES), higher numbers result in elongated simulation durations.

Table 3.3: Mechanism sizes. Depending on the quantity of interest (IDT or LBV), either High Temperature (HT) mechanisms or those including LTC are used.

Mechanism	Type	Count	
		Species	Equations
Gri 3.0	HT	53	325
	/w LTC	n.a.	n.a.
Cai	HT	335	1613
	/w LTC	522	2299
CPN	HT	310	1637
	/w LTC	569	2386
AN	HT	286	1564
	/w LTC	505	2192
Dooley	HT	269	1538
	/w LTC	n.a.	n.a.
UCSD	HT	57	268
	/w LTC	n.a.	n.a.

3.1.1 Gas Composition Calculation

Three of the aforementioned boundary conditions have an impact on the molecular composition of the unburnt mixture: the mixture equivalence ratio ϕ, the mass-based EGR rate w_{EGR}, and the mass based water content w_{H2O}. Additionally, it is also influenced by the formulation of the base fuel. Though, as the fuel mixture definitions have already been given in subsection 5.2 and the whole calculation is done for each mixture individually, it is not treated as an input parameter. This leaves the calculation to the acquisition of the unburnt mixture composition dependent on the equivalence ratio, EGR and water ratio.

To facilitate mixture handling, their definition is given as strings including their base components and molecular fractions; the fractions and the fuel names are divided by underscores. All fuel definitions are stored in simple files and processed by the calculation script. By this, either pure fuels (trivial and redundant) or fuel mixtures like TRF can be defined prior to the calculation itself. For our 50/50 mass-% mixture of TRFE10 with Cyclopentanone, the string would be *"TRF_37.28_Ethanol_10.23_Cyclopentanone_52.49"*. Currently, it is possible to define mixtures of up to three components. If more are desired, the function is easily modified or one of the mixture components (like the see TRF and E10 (TRFE10) in the given example) has to be predefined in the component datafile. Alternatively, the molecular composition can be directly given in a dictionary.

Once the molecular composition of the fuel is known, the unburnt mixture is initialized. Here, the oxidizer composition (per default, air consisting of O_2 and N_2), the desired equivalence ratio, the EGR fraction and composition, and the water rate and composition (e.g., water/ethanol mixtures are possible) are taken into account. The mixture is brought to an equilibrium (or optionally combusted in a reactor; see subsection 3.1.2) and the exhaust gas composition is gathered from there. The mixing of recirculated exhaust gas and fresh mixture is hereby defined by the concept of a stoichiometric EGR ratio (at default settings). This means only inert species are considered for the calculation of this ratio, facilitating the handling for the engine simulations later in the workflow. The concept is further explained in [114]. This process is iterated with the exhaust gas of the previous cycle always being taken for the calculation of the unburnt mixture. It stops, as soon as the deviations between the species fractions of two iteration steps are smaller than a certain threshold. Then, the unburnt mixture composition for this operating point is known, and it can be used in the following calculations. The whole process of finding the mixture definition is also schematically drawn in the flow chart in figure 3.1.

3.1.2 Limited Residency Times

In [114] and [116], the concept of limited residency times for the determination of EGR compositions was first introduced by the author of the current

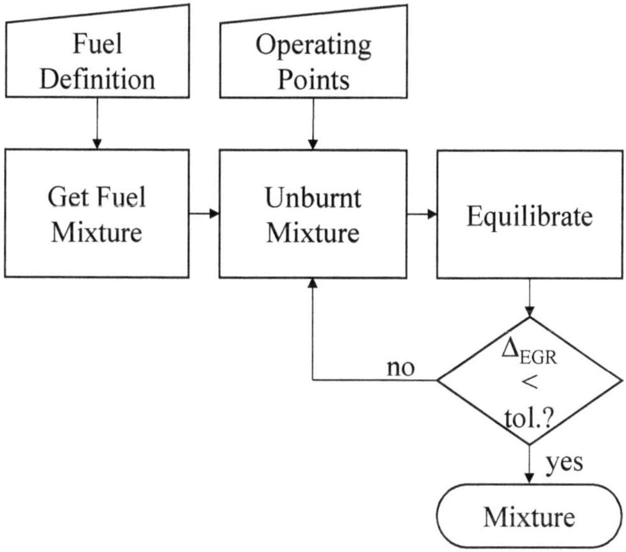

Figure 3.1: Schematic overview for the iterative process of unburnt mixture definition calculation.

work. Instead of assuming instantaneous and complete combustion by taking the equilibrium composition, a more detailed approach is chosen: The fresh, unburnt mixture is put into a (in the case of ignition delay calculations: second) reactor, it gets ignited by a small, hot hydrogen jet, and the reactor state is advanced by 0.05 s, roughly equalling the time between combustion start and Inlet Valve Opens (IVO) of a class C car engine at 1500 min^{-1}. All reactions and species concentrations are then frozen, simulating a rapid cooldown by a fresh mixture. Next, the resulting composition is taken as a definition for the exhaust gas content of the subsequent cycle. This procedure is repeated until the change for each individual species concentration compared to the last cycle is less than a predefined threshold (default: relative 0.02) or the maximum number of iterations has been reached. The mixture formation of the last cycle is then used as EGR composition definition for the proceeding calculations. The principle is further explained in the aforementioned publications [114][116].

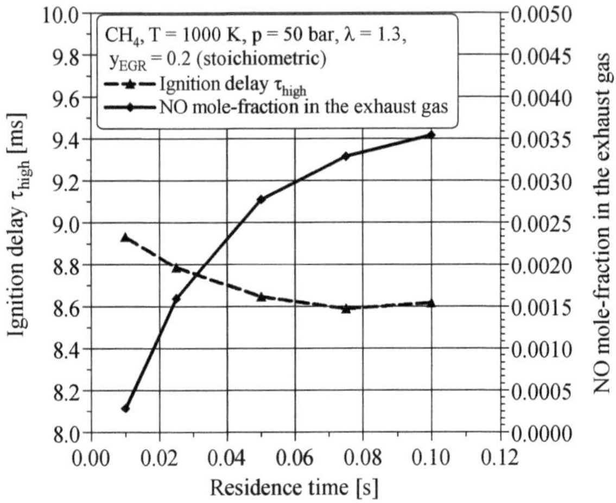

Figure 3.2: NO content and ignition delay dependence from limited residency times for Methane [116].

Since for laminar burning velocities, the effects of additional NO and NO_2 are negligible and the calculation times are increased significantly, mainly due to the less stable equation system with higher amounts of NO_x, the method of considering limited residency times is only used for the calculation of ignition delay times.

3.2 Laminar Flame Calculations

As described in chapter 2, the analysis can be limited to laminar pre-mixed flame propagation. In Cantera, this is resembled by the Free Flame Object. It solves the linear equation system consisting of the conservation equations and the reaction equations. A typical flame profile with temperature and species profiles as well as the division into pre-heating and reaction zones is shown in figure 3.4. It can be seen how reactants are converted into (intermediate) products and how their concentrations change over the profile accordingly. In addition to that, the concentrations of two exemplary species, hydrogen peroxide (H_2O_2), and hydroperoxyl (HO_2) are also depicted.

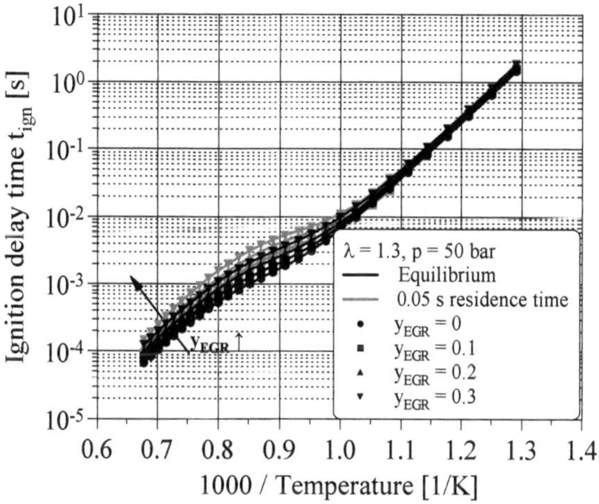

Figure 3.3: Temperature and pressure sweep for ignition delay times of Methane with and without taking limited residency times into account [116].

Both of those molecules are representatives of early combustion stage radicals. They react further to organic hydroperoxides and then break up into hydroxyl radicals ($^{\bullet}OH$).

3.2.1 Laminar Burning Velocity Calculation

A predefined mixture consisting of fuel, air, recirculated exhaust gas, and external (additionally injected) water is burnt in a freely propagating flame. The calculation itself follows the explanations given in section 3.2. The result is given in the form of a class containing the temperature, speed of propagation (also referred to as *laminar burning velocity*), and the concentrations of all species included in the reaction kinetics mechanism for each grid point over the flame width. As the laminar burning velocity is defined for the unburnt fuel, its value is taken from the first point in the grid, ensuring that it is not influenced by any heat release or pressure increase. This also means that the temperature gradient at this point must be negligible. If not, no stable flame has formed within the grid, and the temperatures, even at the first grid point, are already elevated. If this happens during calculation, it is

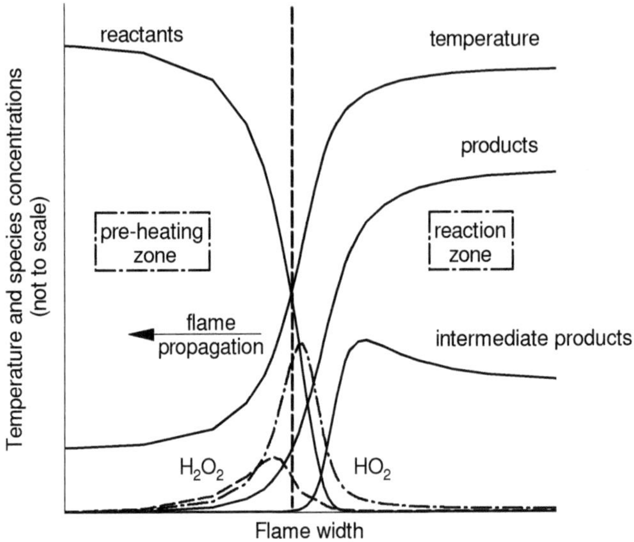

Figure 3.4: Species concentrations and temperature profile for the cross section of a freely propagating laminar flame. The flame profile is divided into two regimes: the pre-heating and the reaction zone. The flame propagates from right to left.

recognized by the script and the operation point is recalculated with slightly adjusted grid parameters.

An exemplary temperature and burning velocity profile over the flame width is shown in figure 3.5. The low or absent gradient in the unburnt array is clearly noticeable. Once the grid is too wide, the accuracy of the extreme temperature rise in the inner layer of the flame gets too low, and the calculation also has to be redone.

Exemplary simulations for the laminar burning velocities of the evaluated neat fuels at an engine reference point of 50 bar and 800 K are shown in figure 3.6. The solid line without markers shows the results for TRF as a reference. As especially the leaner regions are of interest if the combustion efficiency shall be increased, Anisole, Cyclopentanone, and Methanol are attractive future fuels. Apart from Methyl Formate and Dimethyl Carbonate, all fuels show increased burning velocities when compared to the gasoline

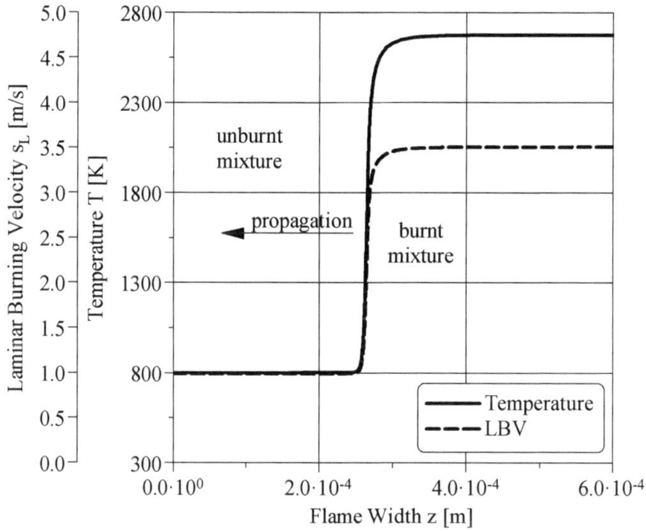

Figure 3.5: Temperature and flame speed profile of a laminar propagating flame.

base fuel. Presumably due to its high oxygen content, methanol shows a better performance for rich equivalence ratios than Anisole or Cyclopentanone.

3.2.2 Laminar Flame Thickness Calculation

Once the solution for the laminar free flame is acquired, all the data needed for the calculation of the laminar flame thickness is available. Since they are simple algebraic calculations, all methods previously mentioned in subsection 2.2.5 can be calculated for each operation point without significantly increasing the calculation time. For the used heat release rate criterion, the point of maximum heat release rate in the flame is looked for, then the thermal conductivity, the specific heat capacity, and the density of the mixture at this point are taken from the free flame calculation result.

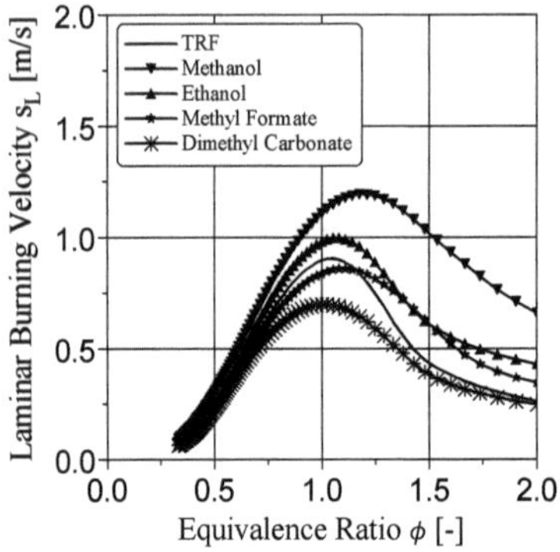

(a) TRF compared to Methanol, Ethanol, Methyl Formate, and Dimethyl Carbonate.

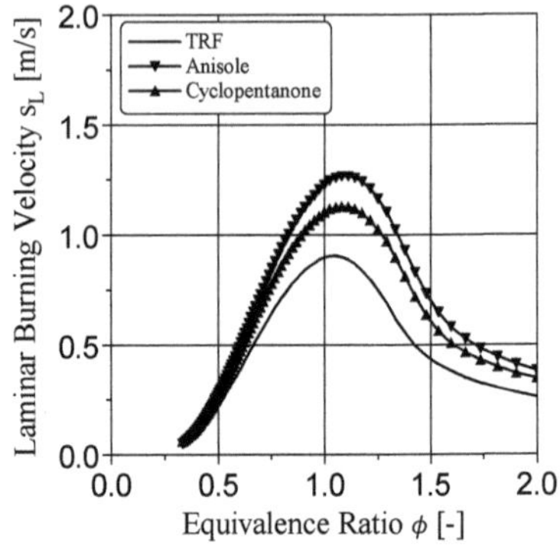

(b) TRF compared to Anisole and Cyclopentanone.

Figure 3.6: Comparison of the laminar burning velocities of the neat fuels Methanol, Ethanol, Methyl Formate, Dimethyl Carbonate, Anisole, and Cyclopentanone with TRF at 50 bar and 800 K.

3.2.3 Lewis Number Calculation

The Lewis number calculation is implemented to be able to determine the ratio between thermal and species diffusion into the flame and how it is affected by changing boundary conditions and fuel. As the calculation is similar to that of the laminar flame thicknesses, it is done for every operation point. With this, even older data can be evaluated later if required.

3.3 Ignition Delay Time Calculation

Contrary to the calculations for the laminar flames, the ignition delay times are acquired from reactor type simulations. Similar to the measurements, the predefined mixture (see subsections 3.1.1 and 3.1.2) is brought into a vessel of constant volume with temperature, pressure, and mixture composition depending on the case that is currently being calculated. The reactor simulation starts and advances until a time limit is reached. Similar to the free flame simulation, species concentrations and a temperature profile are gathered as a result, this time as functions of time. With this, the ignition delay, the time it takes the mixture to self-ignite, can be determined with multiple methods. They either rely on the temperature profile or the species concentrations as input data (see the following enumeration) and are all calculated during runtime. Usually, their quantities lie really close together; depending on the use case (e.g. comparison to measurement data that was acquired by studying the pressure rise in a rapid compression machine), one method may be preferred over the other. More information on the criteria and their use can be found in [114].

(a) **Temperature Criteria**

Temperature rise of 300 K τ_{300K}

Temperature rise of 400 K τ_{400K}

Temperature gradient over 1E7 $\tau_{\nabla T}$

(b) **Species Concentration Criteria**

peak of OH concentration $\tau_{^{\bullet}\mathrm{OH,max}}$

peak of the gradient of OH concentration $\tau_{^{\bullet}\mathrm{OH,grad}}$

3.4 Multithreading Capabilities

For both laminar flame and ignition delay data, a wide array of boundary conditions needs to be rasterized and calculated. Even for low accuracy, with up to 6 variables, this can easily end up in variations of over 100,000 cases for a single fuel. Especially for the free flame calculations that, depending on the mechanism complexity, can take over an hour for a case to converge, a shared calculation on multiple processor cores is preferred over sequential work on a single core.

3.4.1 Multithreading Cantera Applications

At the time of the implementation, there was no explicit support for Cantera calculations in multiprocessing. The author is aware that since September of 2021, there has been an existing multiprocessing example for the simultaneous calculation of species viscosities with the python module *multiprocessing* on the Cantera webpage. The main functions of the Cantera example and their calls are roughly summarized in listing 3.1[1]. The *init_process* function is called once for every process. The *get_thermal_conductivity*-function (not further defined here) is mapped onto the pool. This solution works well for the simple use case in the example, but it has two problems when the complexity of the mapped function is increased: First, the mechanism is called for each process individually. This is done because the main code of Cantera is written in C, and therefore there is no possibility to share the *Cantera.Solution* python class between threads. So every thread has to have its own instance of the same class. But calling the classes *init* function for each thread individually can hinder the scripts' performance, especially for big mechanisms with low temperature chemistry. Secondly, the findings of *Hann* [48] have shown that when doing laminar flame calculations, it can be profitable to cancel the solving process if it takes too much time and restart it with adjusted grid parameters. While the integration of a timeout and a loop was no problem for sequential calculations, it is a bit more problematic for multiprocessing pools.

Listing 3.1: Core elements of the Cantera multiprocessing example.

```
import cantera as ct
import itertools

# Global storage for Cantera Solution objects
gases = {}

def init_process(mech):
    """
    This function is called once for each
    process in the Pool. We use it to
    initialize any Cantera objects we need
    to use.
```

[1] Further can be found on the Cantera website: https://www.cantera.org/examples/python/transport/multiprocessing_viscosity.py.html.

```python
    """
    gases[mech] = ct.Solution(mech)
    gases[mech].transport_model = 'Multi'

def parallel(mech, predicate, nProcs, nTemps):
    """
    Call the function ``predicate`` on ``nProcs``
    processors for ``nTemps`` different temperatures.
    """
    P = ct.one_atm
    X = 'CH4:1.0, O2:1.0, N2:3.76'
    with multiprocessing.Pool(processes=nProcs,
        initializer=init_process,
        initargs=(mech,)
    ) as pool:
        y = pool.map(predicate,
            zip(itertools.repeat(mech),
                np.linspace(300, 900, nTemps),
                itertools.repeat(P),
                itertools.repeat(X)))
        return y

if __name__ == '__main__':
        nPoints = 5000
        nProcs = 4

        parallel('gri30.yaml',
                    get_thermal_conductivity,
                    nProcs, nPoints)
```

3.4.2 Implementation of Timeout Functionality

Two main ideas are pursued to allow for the easy use of Cantera with multiprocessing and optional timeouts. The handling of the *Cantara.Solution* class is improved by loading the mechanism once and instancing it by creating an iterable with a length of n (number of parallel cases), holding an instance of the same class for each thread. This way, all the instancing can happen in the preprocessing. This reduces the loading of data for each thread and decreases overall calculation time. To include the possibility to timeout threads in multiprocessing, the solution proposed by *McFarlane* [82] is adapted to the altered use case: In addition to an *in*- and an *out*-queue that contain the boundary conditions or, in the case of the *out*-queue, additionally all solutions for the cases, a third *worker*-queue is introduced. Additionally, since the use of processes does not lead to the desired results, a second *pool* of only a single worker per case is created, making the thread abortable if desired. As the *multiprocessing.pool* class is not allowed to have children, a second *ThreadPool* is imported from *multiprocessing.dummy*. This approach comes with its own risks, but it worked flawlessly for all applications. An abridged version of the created code can be found in listing 3.2.

Listing 3.2: Proposal for a Cantera multiprocessing aproach with included timeout.

```
import cantera as ct
import multiprocessing as mp

from functools import partial
from multiprocessing.dummy import Pool as
    ThreadPool


def abortable_calculate(func, varianten,
                                   **kwargs):
    """

    Wrapper function to make the "helper"-function
    abortable by timeout.

    Parameters
    ----------
    func : function
      Function to be run with a timeout.
    varianten : Dictionary
      Zipped input data for the calculate function.
    **kwargs : 'timeout': Integer
      Maximum time after which function is aborted.
      Given in seconds.
    **kwargs : 'max_error': Integer
      Maximum number of loop iterations


```

```
    Returns
    -------
    out : Dictionary
      Resultdictionary from wrapped function.

    """
    error = 0
    out = {}

    # get indices and rows again from zipped input
    # data
    index, row = varianten

    timeout = kwargs.get('timeout', None)
    max_error = kwargs.get('max_error', 3)

    p = ThreadPool(1)

    while error < max_error:
      state, out2 = helper(func, varianten, error,
                                   p, timeout)

    if state:
      out = out2
      break

    else:
      print('Calculation timed out. Error: {
            }.'.format(error))
      error += 1
      out['counter'] = error

    return out

def helper(func, varianten, error, p, timeout):
    """

    Call the calculation function.

    Implemented to get around double loops.

```

```
    Parameters
    ----------
    func : Function
      Function to be run with a timeout.
    varianten : Dictionary
      Zipped function input to be forwarded to
      the function. Gets unzipped and zipped
      again to add/refresh the error counter
      in it.
    error : Integer
      Amount of errors occured during function
       handling.
    p : Multiprocessing pool
      Worker pool for multiprocessing (See
      "abortable_calculate" definition).
    timeout : Integer
      Maximum time after which function is
      aborted. Given in seconds.

    Returns
    -------
    bool
      Did the function call end before timing
      out?
    out : Dictionary
      Result dictionary.

    """
    # unpack zipped input:
    index, row = varianten
    er = pd.Series([error], index=['counter'])
    row = pd.concat([row, er])

    # zip again, now containing error
    functionInput = (index, row)

    res = p.apply_async(func, functionInput)

    try:
      out = res.get(timeout)
      state = True
```

```
        except mp.TimeoutError:
            out = 'timed out'
            state = False

        return state, out

    def collectMyResult(result):
    """

    Collect result from worker.

    Gets function result once it is ready. Allows
    for timing out functions. Writing it to the
    global result list.

    Parameters
    ----------
    result : object. Depending on function.
        Result of the function that is called by
        multiprocessing.pool.

    returns
    -------
    None.

    """
    global enum
    global results

    enum += 1
    print('Finished case {} / {}.'.format(
            enum, Cases))
    results.append(result)  # write to global list

def main():
    results = []
    max_time = 1800  # abort after 1800 s
    cores_mp = 16

    pool = mp.Pool(cores_mp, maxtasksperchild=1)

```

```
    # funtionInput is an iterable containing all
    # starting conditions in the form of tuples

    for f in functionInput:
        abortable_func = partial(abortable_calculate,
          calculate, timeout=max_time)
        pool.apply_async(abortable_func, (f, ),
                         callback=collectMyResult)

        pool.close()
        pool.join()

if __name__ == '__main__':
    main()
```

4 Automation of A New Fuel Implementation

Although the introduced models have been used with multiple fuels and validated against measurement data of varying engines, the process of implementing a new fuel mixture is only shown for the two compositions, TRFE10 in a 50 % (m/m) mixture with Anisole ($E10AN_{50m}$) and TRFE10 in a 50 % (m/m) mixture with Cyclopentanone ($E10CPN_{50m}$). They underwent extensive research in the FVV project 1348 *Fuel Composition for CO_2 Reduction* that was running in parallel with this work. The information and measurement data available for those two fuel mixtures makes them ideal for a demonstration of the implementation process. All other fuels and fuel mixtures that have been examined are validated and discussed in chapter 6.

4.1 Schematic Overview

The whole automation process is pictured schematically in figure 4.1. The user provides a fuel definition and a reaction kinetic mechanism and choses the desired output. Depending on this choice, the calculations for the miscellaneous fuel properties, the laminar free flame or in the combustion vessel are started. The dashed line shall symbolize that all those calculations require different amounts of calculation time and that the user is advised to check all calculation results for consistency and plausibility. This also is the reason why the user is forced to start the fit processes by hand. Once they finished, all fuel relevant data is either directly provided (e.g., the lower heating value LHV) or in the form of correlation fits. It can then be stored in a fuel properties database and/or integrated into existing calculation modules.

S. K. Crönert, *A Complete Methodology for the Predictive Simulation of Novel, Single- and Multi-Component Fuel Combustion*, Wissenschaftliche Reihe Fahrzeugtechnik Universität Stuttgart, https://doi.org/10.1007/978-3-658-43075-7_4

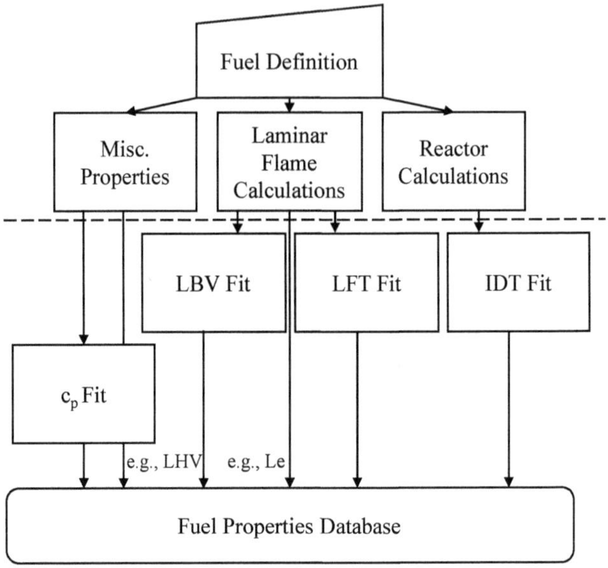

Figure 4.1: Schematic overview of the presented toolchain workflow.

4.2 Fuel Properties

4.2.1 Reaction Kinetics Dependent Properties

All properties dependent on the reaction kinetics (laminar burning velocities, laminar flame thicknesses, ignition delay times) are acquired according to chapter 3. They undergo a quick post-processing to sort out unrealistic data and failed cases. Afterwards, they can be fitted to the correlations (see 2.3.3, 2.3.4, and 2.3.5).

4.2.2 Specific Heat Capacities

The specific heat capacities of the mixtures used are calculated according to eq. 2.16, for a representative temperature range with sufficient resolution

(250 - 2500 K in increments of 10 K). With the generated data for the mixture, a new NASA polynomial is fitted. This way, numeric errors and non-physical miscalculations that could occur with a simple mole-fraction-weighted, linear interpolation between polynomial constants can be bypassed. Although the error is assessed to be of minor quality and big deviations might be rare, this safer method was preferred. For all pure components, usually 7-constant NASA polynomials exist in literature data and the sources of the reaction mechanisms. A later end user, on the other hand, might prefer 9-polynomial constants as they are commonly used in more popular 1D simulation software. For this reason, all data is newly fitted to 9-constant polynomials, even if the fit quality might get slightly worse. The fitting process itself is computed with the *minimize* function of the python module *scipy.optimize* and the *Nelder-Mead* method. First, the coefficients for the specific heat capacities a_{1-7} are adjusted for least squared errors, then the same procedure is applied to the remaining variables b_1 and b_2 for enthalpy and entropy, respectively.

The results for the newly fitted polynomial constants for all used mixtures can be found in Appendix A. The polynomial constants of the neat components are unanimously been taken from the corresponding reaction kinetic mechanisms. Original sources for the species properties are usually given within the mechanisms themselves but have not been cross-checked. For completeness, they are also given in Appendix A.

4.2.3 Liquid Densities

The liquid densities are calculated according to the eq. 2.28 given in subsection 2.1.4. This leads to the values collected in table 4.1. The densities are then stored in a mixture properties sheet as they are only needed for the 1D simulation later on.

4.2.4 Dynamic Viscosities

The dynamic viscosities for all mixtures are calculated using eq. 2.21 and eq. 2.22 in subsection 2.1.4. For the exemplary mixture of Cyclopentanone with gasoline TRFE10 in a 50 % (V/V) mixture with Cyclopentanone ($E10CPN_{50V}$), the dynamic viscosities of the mixture components and the mixture itself are given in 4.2 in dependence on pressure and temperature. The missing sensitivity to pressure changes is due to the simplification of assuming ideal gases. The dynamic viscosity arrays for all mixtures are given in Appendix C.

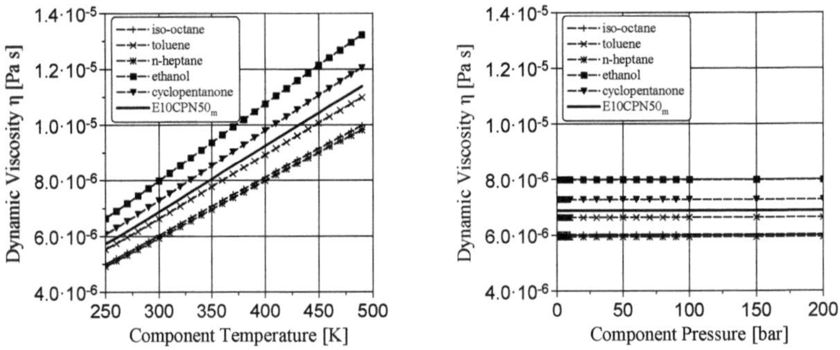

(**a**) Temperature sensitivity (**b**) Pressure sensitivity

Figure 4.2: Dynamic viscosities η of a $E10CPN_{50V}$ mixture and its neat components.

4.2.5 Thermal Conductivities

The thermal conductivities are also calculated using eq. 2.21 and eq. 2.22 in subsection 2.1.4. As the thermal conductivity usually is highly temperature but only marginally pressure dependent - greater influences only occur if a phase change is triggered by the imposed pressure - it is calculated for increments of 50 K at atmospheric pressure. Figure 4.4 depicts the sensitivity of the simulated thermal conductivity of the TRF/E10-mixture to changes in temperature and pressure. Since the pressure influence is not modelled in Cantera, this also means that no effects can be observed for the mixture. As a consequence, the thermal conductivity needed for the engine simulation with any of the modelled fuels is calculated only for constant atmospheric pressure.

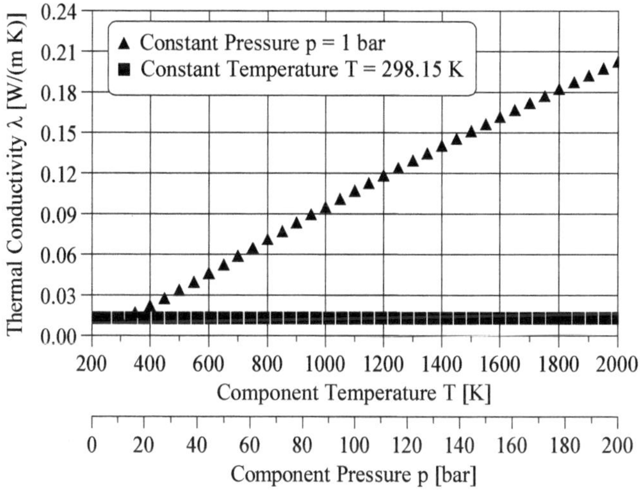

Figure 4.3: Sensitivity to temperature and pressure changes for the thermal conductivity of a TRF/E10-mixture. Simulated with the *Cai* TRF mechanism [14].

For the exemplary mixture of CPN with gasoline $E10CPN_{50V}$ in 4.4, the thermal conductivities of the mixture components are given in dependence on pressure and temperature. The missing sensitivity to pressure changes is due to the simplification of assuming ideal gases. The thermal conductivity arrays for all mixtures are given in Appendix C.

4.2.6 Critical Points

As described in 2.1.4, a simple interpolation is not sufficient to reliably predict the critical point of mixtures. There are multiple options, how an implementation into the calculation script could be done. For this work, eq. 2.23 to eq. 2.27, that rely on the work of *Lee and Kesler* [72] and *Pedersen et al.* [91], are used. They result in the data given in table 4.1.

For further insight, a consolidation of the works of *Moldover* [83], *Heidemann* [51], *Li* [74], and *Stradi* [108] is advised.

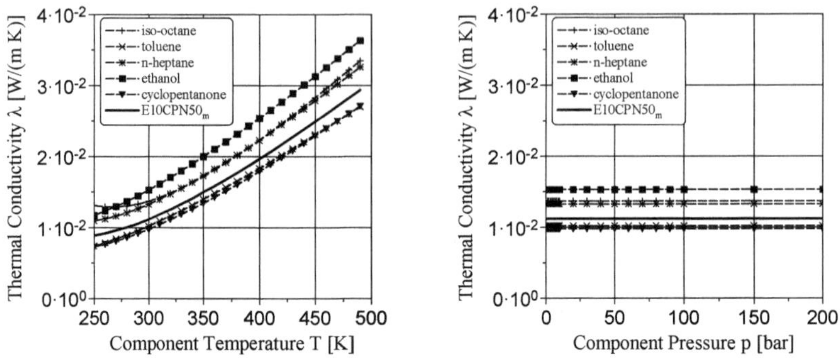

(a) Temperature sensitivity **(b)** Pressure sensitivity

Figure 4.4: Thermal conductivity λ_{th} of a E10CPN$_{50V}$ mixture and its neat components.

Table 4.1: Liquid densities and critical points of the used components and mixtures.

Fuel(-mixture)	ρ_{liq} $kg \cdot m^{-3}$	T_c K	p_c bar	V_c $l \cdot mol^{-1}$
Pure Components				
iso-Octane	0.69	543.9	25.7	0.47
Toluene	0.87	593.0	41.0	0.32
n-Heptane	0.68	540.0	27.4	0.43
Ethanol	0.79	514.0	63.0	0.17
Cyclopentanone	0.95	624.5	46.0	0.27
Anisole	1.00	643.0	42.0	0.34
Dimethyl Carbonate	1.07	557.0	48.0	0.26
Methyl Formate	0.98	487.0	60.0	0.17
Hydrogen	70.0	32.94	12.8	0.06

Continued on next page

Table 4.1: Liquid densities and critical points of the used components and mixtures.(Continued)

Fuel(-mixture)	ρ_{liq} kg·m^{-3}	T_c K	p_c bar	V_c l·mol^{-1}
Gasoline				
TRF	0.73	557.7	29.78	0.41
TRFE5	0.74	554.2	31.82	0.38
TRFE10	0.74	550.9	33.82	0.35
50 % Mixtures				
$E10CPN_{50m}$	0.83	586.3	39.38	0.31
$E10AN_{50m}$	0.85	592.1	37.40	0.35
$E10CPN_{50V}$	0.85	590.9	40.14	0.30
$E10AN_{50V}$	0.87	599.0	38.01	0.35
EN228				
E5 CPN	0.75	560.9	33.03	0.37
E5 AN	0.76	564.3	32.91	0.38
5 wt.-% Oxygen				
E10 CPN	0.75	556.2	34.61	0.35
E10 AN	0.76	539.0	34.46	0.35
Misc.				
$DMC_{65}MF_{35V}$	1.04	530.0	52.5	0.22

4.3 Laminar Flame Speed Correlation Fit

The use of the flame speed correlation developed by *Mueller* [84], modified by *Ewald* [33], and extended by *Hann* [48] and the author is severely limited by the immense effort that has to go into fitting the complex equation structure to the reaction kinetic calculation results. The workflow is not feasible fo the use with a wider range of fuels, if for the integration of each new fuel an experienced person has to invest several days to get a good correlation agreement. To alleviate this, an automated fit procedure has been created that returns reliable, reproducible results every time and speeds up the tool chain use. Additionally, the user does not have to supervise the operation but can concentrate on more demanding work instead. The fitting script handles and cleans up the reaction kinetic calculation data and fits eq. 2.66 to eq. 2.72 to it. The process is split into multiple smaller optimizations as only those regions affected by a certain correlation parameter are recalculated and compared to their respective target values. This way, the optimization time is drastically reduced.

Once this optimization is done, the deviation can be plotted in a colormap. Figure 4.5 shows this visualization for a few selected operation points with the base fuel Gasoline with 10 % (V/V) Ethanol Content (E10). Lighter shades show good correlation between fit and reaction kinetic data, while darker shades show the points of higher deviations. Additionally, a pressure/temperature trace from Engine A (5.1) is integrated into the figures to show the region where good agreement is most important. It shows the pressure and temperature combinations during combustion (5 - 95 % MFB) for an operating point of 15 bar load and 2000 min^{-1} engine speed. For lower loads and engine speeds, the p/T trace will be offset towards the lower left corner (lower temperatures and pressures). When looking at the error plots, it can be seen, that the base correlation has really good agreement for the whole range of pressures and temperatures. With additional cross-influences (enleanment by the means of overstoichiometric conditions - see figure 4.5b - or increased EGR ratios - see figure 4.5c), the model quality is lower. Especially for extreme conditions at the lower limits of the fitting range, higher errors occur. Although those deviations might seem massive, they are on the one hand corresponding to small absolute burning velocities close to the lean burn limit, on the other hand they lie far outside the engine relevant boundary conditions. The exemplary look at the fits for the base fuel shows what to expect for the newly included fuels and fuel mixtures. The aim is to achieve at least the same fit quality for their laminar burning velocities as with gasoline.

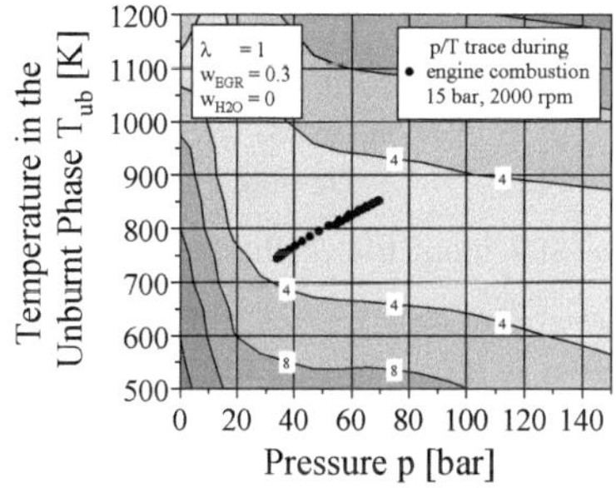

(**a**) Stoichiometric conditions.

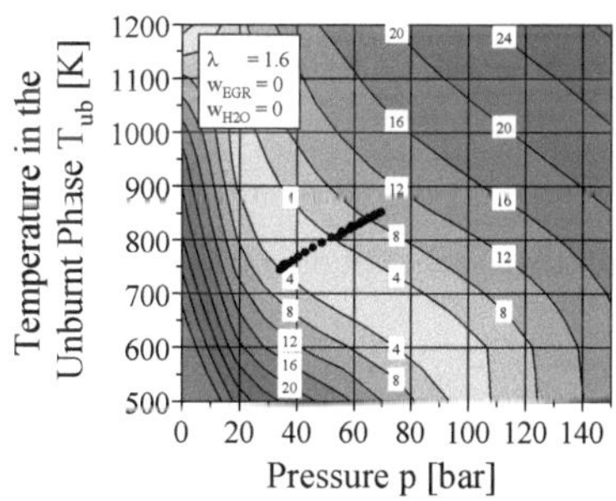

(**b**) Lean conditions (λ = 1.6).

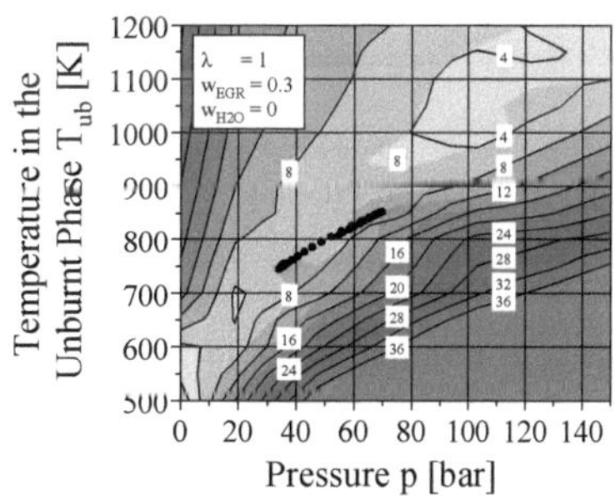

(**c**) Extended EGR ratio (w_{EGR} = 0.3).

Figure 4.5: Laminar burning velocity fit quality for TRFE10 with varying boundary conditions. The colorscale shows the relative error between reaction kinetic calculation results and fit values. Darker colors correspond with higher deviations.

Other than with the ignition delay times, the laminar burning velocities are explicitly needed for the pure fuels. The quality of their correlation fits will be discussed first. The comparison for neat Anisole in figure 4.6 shows great quality in the relevant regimes. Even for higher dilutions (now λ = 2.0) and higher EGR rates, no major flaws are observed.

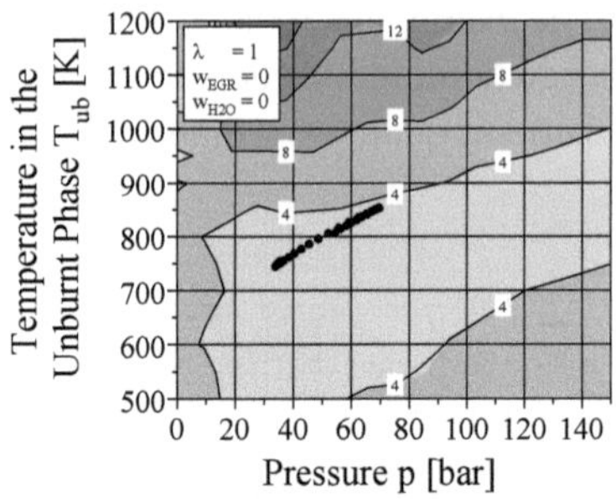

(a) Stoichiometric conditions.

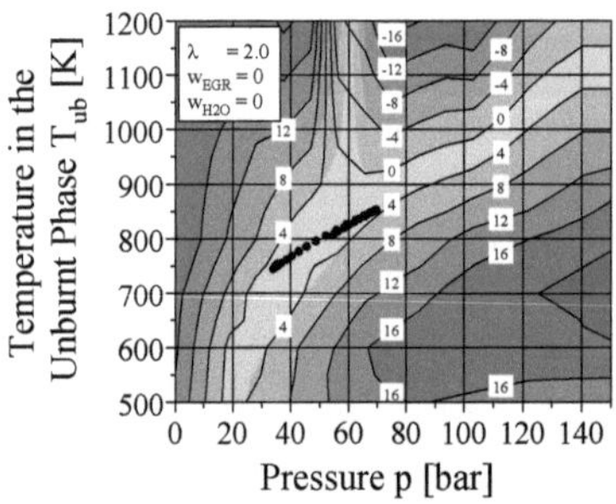

(b) Lean conditions (λ = 2.0).

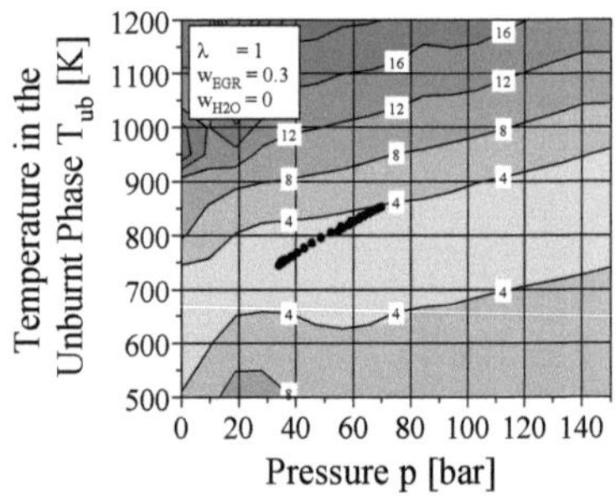

(c) Extended EGR ratio (w_{EGR} = 0.3).

Figure 4.6: Laminar burning velocity fit quality for Anisole with varying boundary conditions.

For neat Cyclopentanone, the same variation is shown in figure 4.7. While the fit quality also fulfils the requirements set, the errors at lower temperatures are more similar to those observed with the base fuel.

The correlation errors for the 50/50 V/V mixtures show acceptable error ranges as well (see figures 4.8 and 4.9). The only major deviations occur in regions where commonly no engine combustion takes place. All laminar flame speed correlations are within the demanded quality range and can be used for predictive combustion simulation.

Exemplary correlation quality plots for the laminar burning velocities of other fuels and fuel mixtures can be found in the Appendix D.

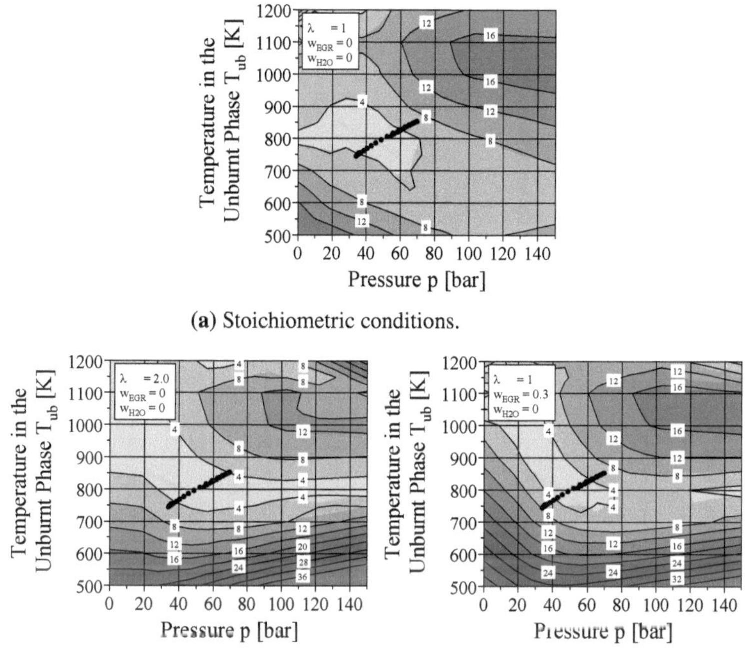

(**a**) Stoichiometric conditions.

(**b**) Lean conditions (λ = 2.0).

(**c**) Extended EGR ratio (w_{EGR} = 0.3).

Figure 4.7: Laminar burning velocity fit quality for Cyclopentanone with varying boundary conditions.

4.4 Laminar Flame Thickness Correlation Fit

All previously calculated laminar flame thicknesses (see section 3.2.2) are fit to eq. 2.73. As the additional data (flame speeds and mixture densities) has been calculated simultaneously, the whole correlation fit comes down to minimizing the error between calculated flame thicknesses and model output by calibrating the pre-factor $\alpha_{\delta_{\text{L}}}$. This is done with the Python method *optimize.minimize* from the *scipy* module. For the minimization, a *Nelder-Mead* algorithm is used.

Once the calculation is done, the results can again be compared to the calculated flame thickness values. For the TRF base fuel, this results in figure

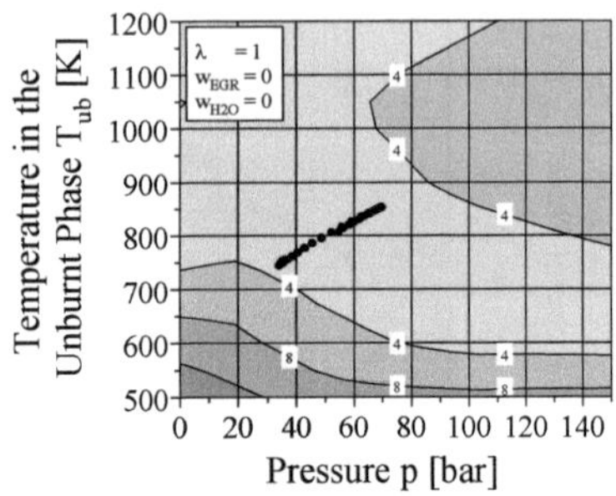

(**a**) Stoichiometric conditions.

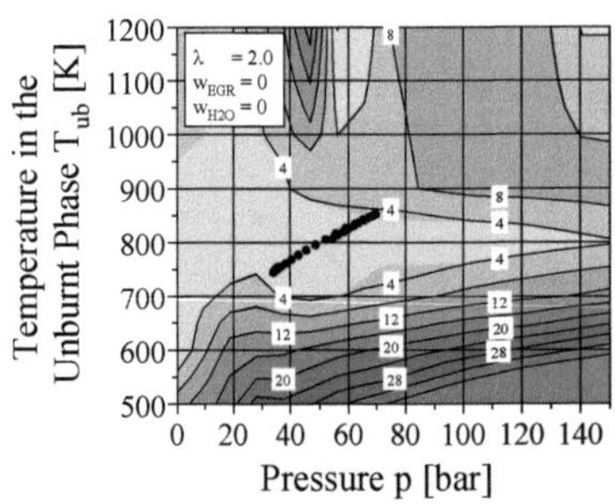

(**b**) Lean conditions (λ = 2.0).

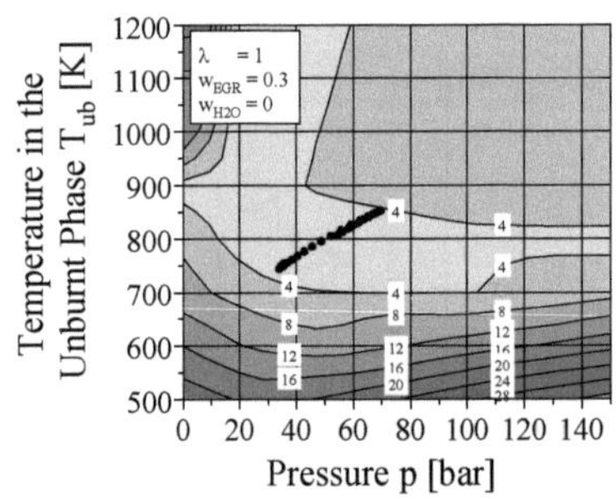

(**c**) Extended EGR ratio (w_{EGR} = 0.3).

Figure 4.8: Laminar burning velocity fit quality for $E10AN_{50V}$ with varying boundary conditions.

4.10. The relative deviation is given by the iso-lines and the colormap, with lighter shades representing better agreement between the correlation and the calculation results. Overall, it can be seen that the base fit has good quality within the observed range of boundary conditions and the assumptions made by *Hann* [48] are valid. The maximum deviation lies at 3 %.

If the same principle is applied to the pure components Anisole and Cyclopentanone and their 50/50 V/V mixtures with the base fuel, a similar correlation quality can be achieved. Especially for neat Cyclopentanone, the fit seems to perfectly match the calculation results.

For the 50/50 V/V mixtures, the quality is again as good as with the base fuel. There are no major trends recognizable that would need any kind of further

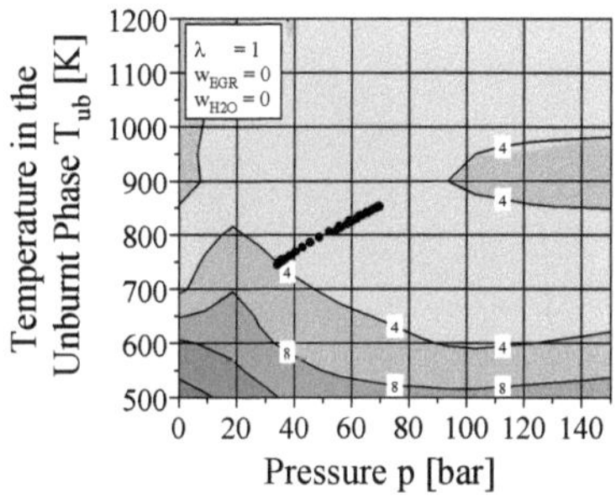

(**a**) Stoichiometric conditions.

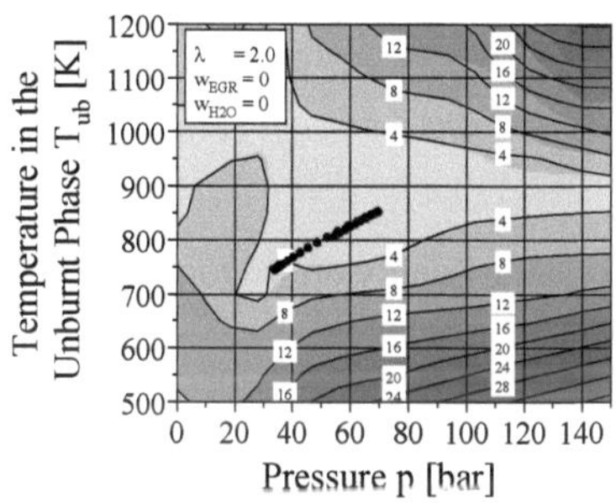

(**b**) Lean conditions (λ = 2.0).

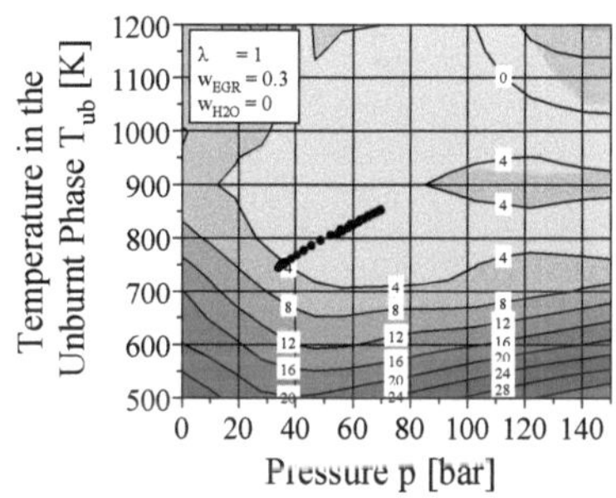

(**c**) Extended EGR ratio (w_{EGR} = 0.3).

Figure 4.9: Laminar burning velocity fit quality for $E10CPN_{50V}$ with varying boundary conditions.

adjustments. In summary, the flame thickness model is easily applicable to newer or less common fuels, and the fitting process is adequate for the quality required for practical use.

4.5 Universal Auto-Ignition Correlation

While for the neat fuels Anisole and Cyclopentanone, according to the observations made in section 3.3, a single *Arrhenius* equation to model ignition delays is sufficient and no two-stage auto-ignition can be observed, this is different for the $E10AN_{50m}$ and $E10CPN_{50m}$ fuel mixtures.

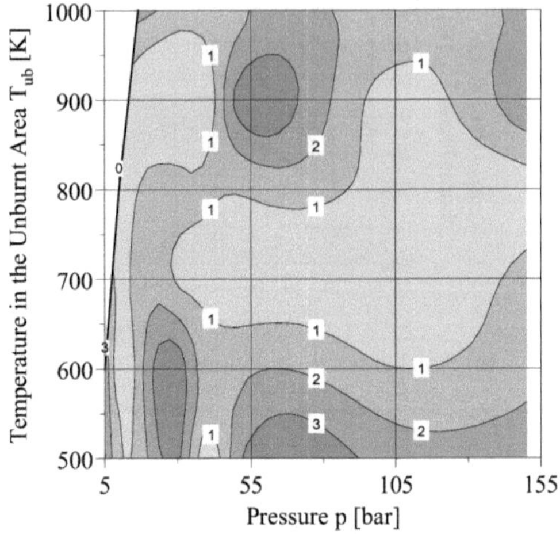

Figure 4.10: Laminar flame thickness correlation quality drawn for the base fuel TRF. Pressure and temperature variation at $\lambda = 1$.

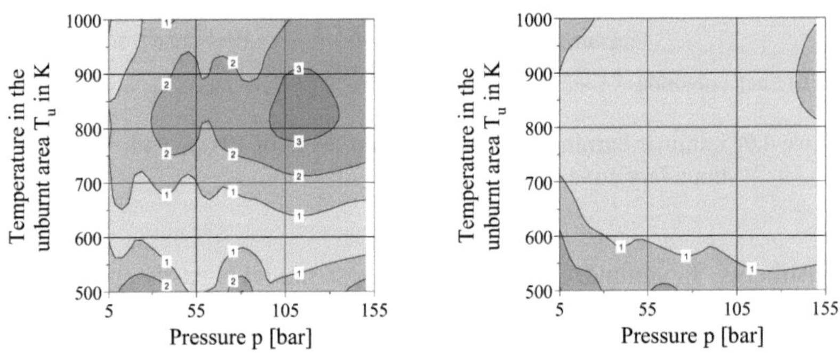

(a) Anisole **(b)** Cyclopentanone

Figure 4.11: Laminar flame thickness correlation quality drawn for the pure components. Pressure and temperature variation at $\lambda = 1$.

The *Fandakov* ignition delay model referred to in 2.3.2 has been built to fit reaction kinetics calculation data to multiple qualities of gasoline fuel with varying ethanol content [34]. In its final version, it consists of 3 sets of equations, including 504 correlation coefficients. An extension of this model for

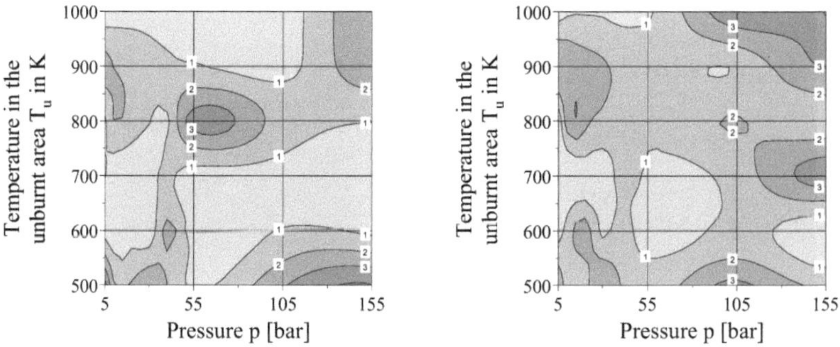

(a) E10AN50v

(b) E10CPN50v

Figure 4.12: Laminar flame thickness correlation quality drawn for the 50 % V/V mixtures. Pressure and temperature variation at $\lambda = 1$.

even more additives is not justifiable, and even fitting a new fuel or fuel mixture to the existing correlation comes with great effort. Because of this, the base idea of using multiple, nested equations to store the coefficients of a triple-Arrhenius approach is kept. The idea is to remove all mixture-related terms and speed up the fitting process, so that any new fuel can be implemented in a matter of minutes. Additionally, the fitting process is automated to later allow inexperienced users to create their own model fit while ensuring a high model quality. With this streamlined and sped-up process, it is easy to acquire new fit coefficients for changing fuel qualities.

This way, the model does not have to account for changing compositions of the base fuel. With this eq. 2.77 and eq. 2.78 are reduced to eq. 4.1 and eq. 4.2. For a slightly better overview, the naming convention of the pre-factors has been adjusted to be in alphabetical order. This means P and Q will now be the pre-factors for the composition based on the equivalence ratio and R and S for the EGR ratio. The indices then distinguish between either the pre-factor A or the exponent B.

$$A_{\mathrm{i,high}} = exp(P_{\mathrm{A,\,i}} \cdot (\lambda - 1)^2 + Q_{\mathrm{A,\,i}} \cdot (\lambda - 1) + R_{\mathrm{A,i}} \cdot (\frac{m_{\mathrm{EGR}}}{m_{\mathrm{cyl}}} \cdot 100)^2 + S_{\mathrm{A,\,i}} \cdot (\frac{m_{\mathrm{EGR}}}{m_{\mathrm{cyl}}} \cdot 100) + U_{\mathrm{A,i}}) \quad \text{eq. 4.1}$$

$$B_{\mathrm{i,high}} = P_{\mathrm{B,i}} \cdot (\lambda - 1)^2 + Q_{\mathrm{B,i}} \cdot (\lambda - 1) + R_{\mathrm{B,i}} \cdot (\frac{m_{\mathrm{EGR}}}{m_{\mathrm{cyl}}} \cdot 100)^2 + S_{\mathrm{B,i}} \cdot (\frac{m_{\mathrm{EGR}}}{m_{\mathrm{cyl}}} \cdot 100) + U_{\mathrm{B,i}} \quad \text{eq. 4.2}$$

To better account for the greater variability in the sensitivities to changing boundary conditions of various fuels, the degree of the polynomials behind P, Q, R, S (all 3rd grade polynomials), and U (formerly exponential functions, now a 4th grade polynomial for high temperature ignition delay and a 3rd grade polynomial for the low temperature ignition delay) is raised. For the temperature increase caused by the first stage of ignition, the same principle is applied. The surrogate-dependent terms are dropped, and while this time the types of polynomials for the pre-factors P to S stay the same, the fitting of the term U is again changed from two coupled exponential functions to a 3rd grade polynomial.

$$C_{\mathrm{i}} = P_{\mathrm{C,\,i}} \cdot (\lambda - 1)^2 + Q_{\mathrm{C,\,i}} \cdot (\lambda - 1) + R_{\mathrm{C,i}} \cdot (\frac{m_{\mathrm{EGR}}}{m_{\mathrm{cyl}}} \cdot 100)^2 + S_{\mathrm{C,\,i}} \cdot (\frac{m_{\mathrm{EGR}}}{m_{\mathrm{cyl}}} \cdot 100) + U_{C,\mathrm{i}} \quad \text{eq. 4.3}$$

Even with this raising the number of correlation coefficients, their count was almost halved to a total of 286 per fuel or fuel mixture. The coefficients for the two fuel mixtures, TRFE10 in a 50 % (V/V) mixture with Anisole ($E10AN_{50V}$) and $E10CPN_{50V}$ can be found in Appendix F.

4.6 Auto-Ignition Correlation Fit

As with the correlation itself, its fit is also divided into three parts: First of all, the high temperature ignition delays are treated. Then, if there are any occurrences of two-stage ignition in the observed operation area, the two other correlations are also used.

4.6.1 High Temperature Ignition Delay Correlation Fit

As they are needed for every fuel, the high temperature ignition delays are fitted first. Due to their complex shape for temperature sweeps, they also take the most effort to achieve adequate results. The algorithm starts by finding the best coefficients for $U_{\mathrm{A,i}}$ and $U_{\mathrm{B,i}}$ for each pressure at stoichiometric conditions without recirculated exhaust gas. This means eq. 2.76 is adjusted to the calculation results, with $A_{\mathrm{i,high}}$ being replaced by $U_{\mathrm{A,i}}$ and $B_{\mathrm{i,high}}$ by $U_{\mathrm{B,i}}$. This leads to eq. 4.4 being solved for each pressure individually. As the area with NTC does not necessarily lie in the same temperature range for each fuel, the location of the transitions between the different zones is found by looking at the second derivation of the ignition delay data. This way, the fit quality is improved, and since no manual input is needed, there is less potential for error.

$$\tau_{\mathrm{i,high}} = U_{\mathrm{A,i}} e^{\left(\frac{U_{\mathrm{B,i}}}{RT}\right)} \qquad \text{eq. 4.4}$$

With the acquired optimum values of $U_{\mathrm{A,i}}$ and $U_{\mathrm{B,i}}$ for each pressure, they themselves are then exchanged for pressure dependent polynomials, and only the polynomial coefficients are used for the correlation. This has the big advantage that interpolation between pressure points is already included when later calling the function to get ignition delay data and does not add to the calculation time. Espccially when empirically fitting polynomials of higher grades to physical data, one has to be really careful and keep the limits of this approach in mind. As they might behave in an unexpected manner outside of the fitting range, it is best to prevent every extrapolation. Once the pressure sensitivity is met within reasonable error margins, the same procedure is done with the λ and the EGR rate variation. For these, there are two slight adjustments to the process. Since the base variation at stoichiometric conditions and without EGR is already covered by the first step and all other influences are covered by adding their respective terms, U_A and U_B are now subtracted from the optimum A_i and B_i before the pre-factors are fitted (eq. 4.5). Additionally, as eq. 4.1 suggests, the λ or EGR variation is fitted to a λ-dependent polynomial of second order and the polynomial coefficients P and Q are themselves fitted to pressure-dependent polynomials (the same obviously applies for the EGR variation). This structure requires extreme caution in the fitting process.

$$P_{A,i,opt} = A_{i,opt} - U_{A,i} \qquad \text{eq. 4.5}$$

Once the base calibration for the pressure dependence and the two for the cross-influences p/λ and p/EGR are done, the results for the $E10AN_{50V}$ mixture are as plotted in the figures 4.13 (a) to (c). Not only are the trends of the reaction kinetic calculations met, but the overall agreement of the fit is also sufficient. For the medium (relative to ICE boundary conditions) pressure range around 50 bar, there seems to be a small oscillation in the calculation results that is not met by the correlation fit. As the reaction kinetic calculations were made with real gas (as opposed to with ideal gas simplifications), this could be caused by the numerical solving process. With the mathematical construction of the triple-*Arrhenius* approach, this curve shape can never be met and thus these errors have to be accepted. Since the error of the ignition delay measurements (that are the basis for the reaction mechanisms used - see 2.2) roughly lies in the range of +/- 50 %, these low deviations will have way less impact on the predictive capability of the simulations than the mechanism creation process itself.

If the same comparison between calculation results and correlation data is made for the $E10CPN_{50V}$ fuel mixture, this results in figure 4.14. The same deviations for medium-range temperatures are also found with the Cyclopentanone mixture.

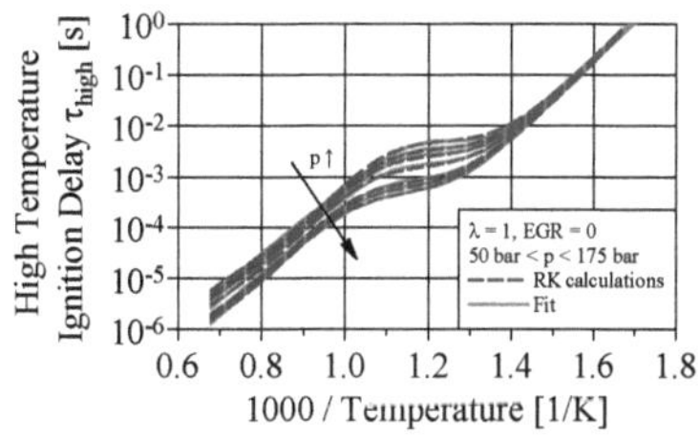

(**a**) Pressure variation fit at stoichiometric conditions and without EGR.

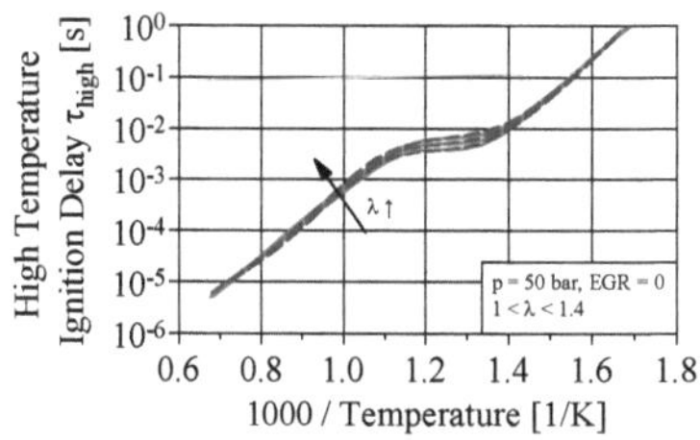

(**b**) Equivalence ratio variation fit at constant pressure of 50 bar and without EGR.

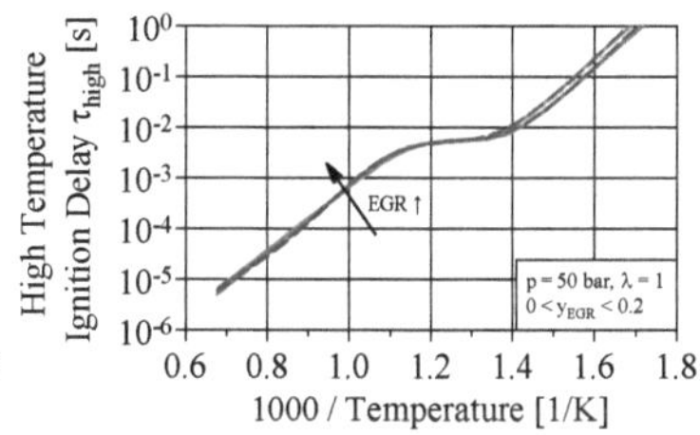

(**c**) Exhaust gas ratio variation fit at constant pressure of 50 bar and at stoichiometric conditions.

Figure 4.13: High temperature ignition delay τ_{high} fit for E10AN$_{50\text{V}}$ with varying boundary conditions.

4.6.2 Low Temperature Ignition Delay Correlation Fit

The low temperature ignition delays obviously can only be fit to data for which two-stage ignition occurs. Prior to the calculation, all other points are dropped. The fitting process then is exactly the same as with the high temperature ignition delay, apart from the fact that only two temperature ranges are fitted (see eq. 2.81). For the E10AN$_{50\text{V}}$ mixture, this results in the correlation fits in figure 4.15. While the overall fit quality is great, for higher temperatures the ignition delay times are slightly overpredicted. Consistent with *Fandakov* is the low sensitivity to changes in the equivalence ratio. Additionally, it can be seen, that as soon as EGR is added, the two-stage ignition is suppressed. It seems like the Anisole share in the mixture is inhibiting low temperature reactions to the point where some additional inert mass in the mixture is enough to stop low temperature pre-reactions completely.

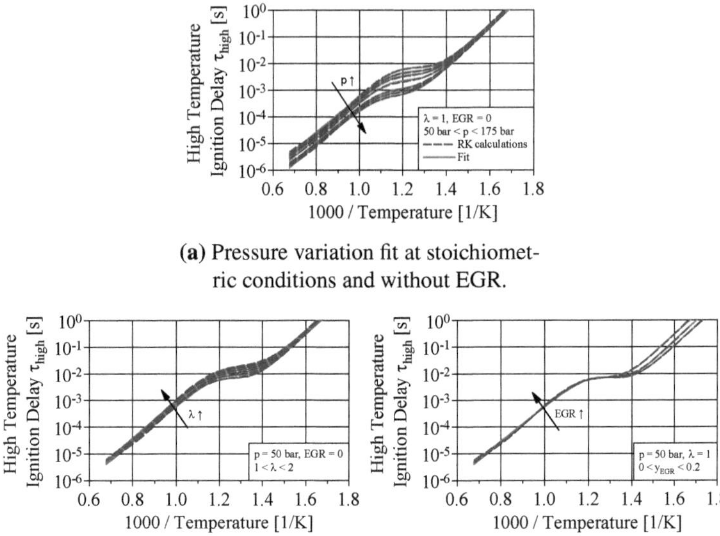

(a) Pressure variation fit at stoichiometric conditions and without EGR.

(b) Equivalence ratio variation fit at constant pressure of 50 bar and without EGR.

(c) Exhaust gas ratio variation fit at constant pressure of 50 bar and at stoichiometric conditions.

Figure 4.14: High temperature ignition delay τ_{high} fit for $\text{E10CPN}_{50\text{V}}$ with varying boundary conditions.

Apart from minor offsets or slightly altered slopes, the low temperature ignition delays for $\text{E10CPN}_{50\text{V}}$ are of the same quality (figure 4.16).

4.6.3 Temperature Increase Correlation Fit

Again, the temperature increase correlation can only be fit to data for that two-stage ignition occurs. First, the fourth grade polynomial with the coefficients C_1 to C_5 is fit for each pressure individually. The pressure dependence of the optimal values for C is then mirrored with U_i. Once the pressure variation quality is good, the procedure follows the one for the ignition delay fit process: For the equivalence ratio and EGR ratio influences, the base value of U_i is subtracted from the optimal values of C_i and the remaining term is then fitted to S and P (or in the case of the EGR variation, Q and R).

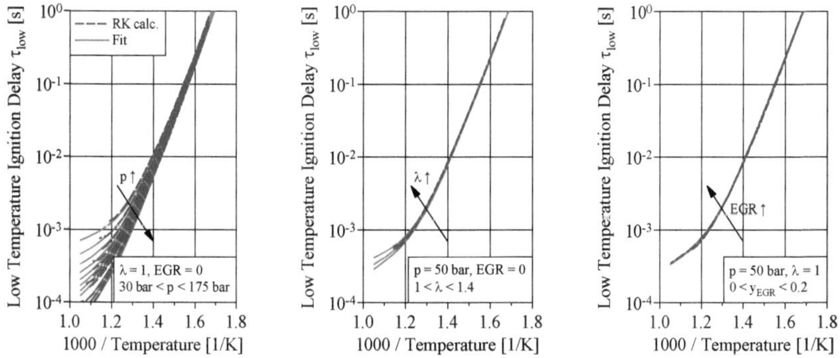

(a) Pressure variation fit at stoichiometric conditions and without EGR.

(b) Equivalence ratio variation fit at constant pressure of 50 bar and without EGR.

(c) Exhaust gas ratio variation fit at constant pressure of 50 bar and at stoichiometric conditions.

Figure 4.15: Low temperature ignition delay τ_{low} fit for E10AN$_{50V}$ with varying boundary conditions.

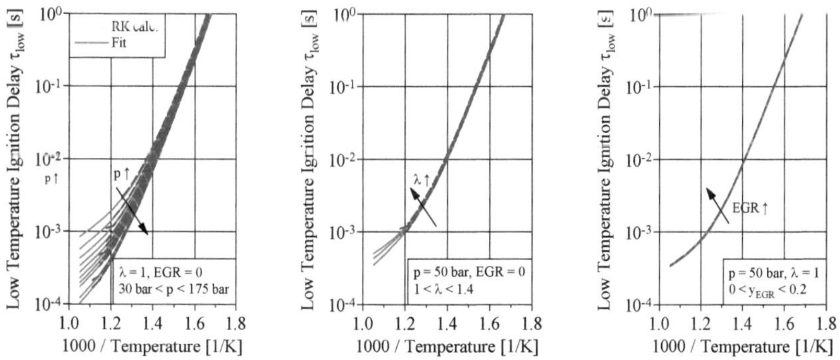

(a) Pressure variation fit at stoichiometric conditions and without EGR.

(b) Equivalence ratio variation fit at constant pressure of 50 bar and without EGR.

(c) Exhaust gas ratio variation fit at constant pressure of 50 bar and at stoichiometric conditions.

Figure 4.16: Low temperature ignition delay τ_{low} fit for E10CPN$_{50V}$ with varying boundary conditions.

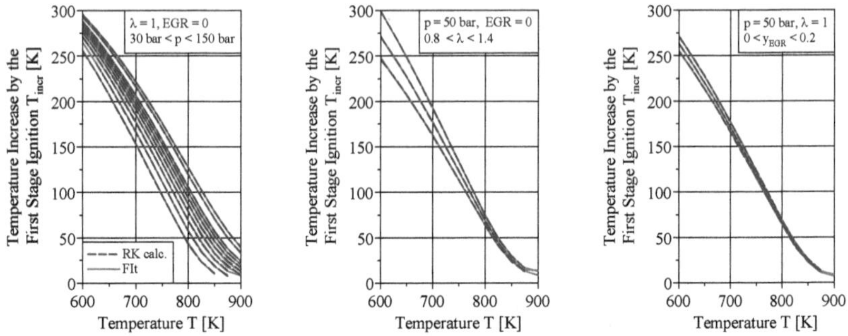

(a) Pressure variation fit at stoichiometric conditions and without EGR.

(b) Equivalence ratio variation fit at constant pressure of 50 bar and without EGR.

(c) Exhaust gas ratio variation fit at constant pressure of 50 bar and at stoichiometric conditions.

Figure 4.17: Correlation fit for the temperature increase by the first stage of ignition T_{incr} for $\text{E10AN}_{50\text{V}}$ with varying boundary conditions.

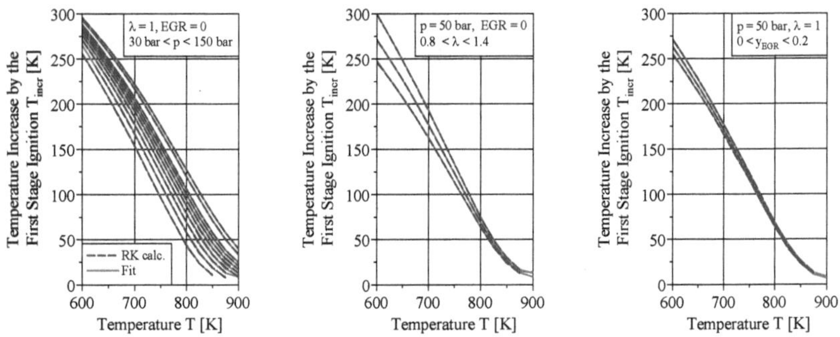

(a) Pressure variation fit at stoichiometric conditions and without EGR.

(b) Equivalence ratio variation fit at constant pressure of 50 bar and without EGR.

(c) Exhaust gas ratio variation fit at constant pressure of 50 bar and at stoichiometric conditions.

Figure 4.18: Correlation fit for the temperature increase by the first stage of ignition T_{incr} for $\text{E10CPN}_{50\text{V}}$ with varying boundary conditions.

4.7 Implementation into 0D Calculation

To allow for flexibility with the integration into the 0D process calculation, the knock model by *Fandakov* was split into two parts. The auto-ignition module and the knock occurrence criterion. This way, they both can be used independently. The depiction in figure 4.19 can be read as follows: for each row, one of the modules can be chosen, depending on personal preferences and experiences. The laminar burning velocity is needed for all turbulent combustion modules. Additionally, there are other possibilities than the presented correlation (e.g., *Gülder* or *Heywood* correlations or tabulated data) not shown in the schematic. As the laminar flame thickness is only needed for some burn rate models, its also absent.

The *Fandakov* knock model has been separated into an ignition delay correlation with auto-ignition determination and the knock criterion itself. This way, they both can be used independently. If a user now wants to calculated knocking combustion of a fuel, depending on the mixture he can chose the IDT correlation that fits the use case the best. For pure gasoline it might be promising to exactly match the surrogate composition with the *Fandakov* ignition delay correlation, for other fuel mixtures that show a distinct NTC behavior in their IDTs, the new correlation should be preferred, and for pure fuels with IDTs close to simple *Arrhenius* straights, a single *Arrhenius*-type fit can be used. The choice of the knock occurrence criterion is then again completely detached from the previous row and the user can chose whatever model fits the needs the best. This also ensures modularity for new models or extensions to existing ones as only the relevant submodule has to be adjusted.

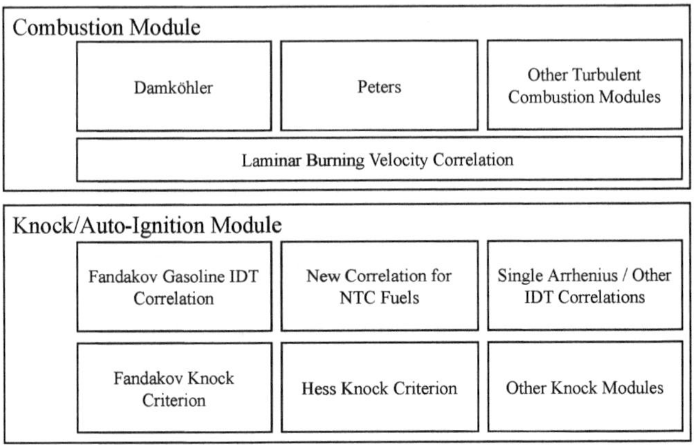

Figure 4.19: Modular integration of the adjusted models into engine combustion simulation and the resulting interchangeability between submodules.

5 Engine Properties and Fuel Compositions

To allow for a better understanding of the diverse validation range, not only in terms of fuels but also for engine types, the used engines and fuels are introduced. Also, the fuel compositions are explicitly stated, so all calculation results can be reconstructed if needed. Additionally, the abbreviations and characteristics of the used fuel compositions are mentioned and summarized.

5.1 Engine and Test Bench Properties

Engine A (Gasoline/Anisole/Cyclopentanone)

The engine test bench measurements for the combustion of the gasoline/Anisole and gasoline/cyclopentanone fuel mixtures have been done at the Chair of Thermodynamics of Mobile Energy Conversion Systems (TME) Aachen on a single cylinder research engine. Its properties are given in table 5.1. With this engine, the charge air is preconditioned by the test bench, and the back pressure is controlled by a valve. For higher loads, the back pressure is controlled to match the intake pressure. This way, a turbine with perfect efficiency is simulated.

Engine B (Methanol)

With Engine B, combustion simulations with methanol have been made during the FVV research project #1307 *ICE2025+: Ultimate System Efficiency*. Its specifications can be found in table 5.2 [62].

S. K. Crönert, *A Complete Methodology for the Predictive Simulation of Novel, Single- and Multi-Component Fuel Combustion*, Wissenschaftliche Reihe Fahrzeugtechnik Universität Stuttgart, https://doi.org/10.1007/978-3-658-43075-7_5

Table 5.1: Engine characteristics Engine A (Gasoline / Anisole / Cyclopentanone).

Property	Symbol	Value	Unit
Cylinder Count	n	1	
Bore	B_{E}	0.075	m
Stroke	s_E	0.113	m
Stroke/Bore Ratio	s_E/B_{E}	1.5	
Displacement	V_h	500	cm^3
Compression Ratio	CR	10.8	$\frac{\mathrm{m}^3}{\mathrm{m}^3}$
Injection		DI	
Aspiration		pre-conditioned	
Exhaust		back pressure valve	

Table 5.2: Engine characteristics Engine B (Methanol).

Property	Symbol	Value	Unit
Cylinder Count	n	4	
Bore	B_{E}	0.0745	m
Stroke	s_E	0.0859	m
Stroke/Bore Ratio	s_E/B_{E}	1.15	
Displacement	V_h	375	cm^3
Compression Ratio	CR	Variable	
Injection		DI	
Aspiration		turbocharged	
Exhaust		VTG	

Engine C & D ($\mathbf{DMC_{65}MF_{35}}_V$)

With engines C and D, $DMC_{65}MF_{35V}$ combustion has been calculated and compared to the measurements made in the *NaMoSyn* research project funded by the German Federal Ministry of Education and Research (BMBF).

Engine C is a single cylinder research engine based on the EA 888 GEN2 engine geometry (table 5.3), engine D is a four cylinder engine based on the EA 888 EVO4 geometry (table 5.4).

Table 5.3: Engine characteristics Engine C (DMC / MF) [5].

Property	Symbol	Value	Unit
Cylinder Count	n	1	
Bore	B_{E}	0.08251	m
Stroke	s_E	0.0866	m
Stroke/Bore Ratio	s_E/B_{E}	1.05	
Displacement	V_h	463	cm^3
Compression Ratio	CR	10.92	$\frac{\mathrm{m}^3}{\mathrm{m}^3}$
Injection		DI	
Aspiration		turbocharged	
Exhaust		VTG	

Table 5.4: Engine characteristics Engine D (DMC/MF) [5].

Property	Symbol	Value	Unit
Cylinder Count	n	4	
Bore	B_{E}	0.0825	m
Stroke	s_E	0.0928	m
Stroke/Bore Ratio	s_E/B_{E}	1.125	
Displacement	V_h	496	cm^3
Compression Ratio	CR	9.6	$\frac{\mathrm{m}^3}{\mathrm{m}^3}$
Injection		DI	
Aspiration		turbocharged	
Exhaust		VTG	

Engine E & F (Hydrogen)

The methodology has also been used to simulate hydrogen combustion on engines E and F. The former is based on a MTU BR2000 (table 5.5) and has been used to predict hydrogen combustion in a cooperation of the Research Institute for Automotive Engineering and Powertrain Systems Stuttgart (FKFS) with Bosch [64]. The foundation for the latter is the Deutz diesel engine TCD 7.8 that was based on a diesel engine with similar geometry (TCD 5.0) [87]. Its specifications are shown in table 5.6.

Table 5.5: Engine characteristics Engine E (Hydrogen) [64].

Property	Symbol	Value	Unit
Cylinder Count	n	12	
Bore	B_{E}	0.150	m
Stroke	s_E	0.130	m
Stroke/Bore Ratio	s_E/B_{E}	0.87	
Displacement	V_h	1984	cm^3
Compression Ratio	CR	11.5	$\frac{\mathrm{m}^3}{\mathrm{m}^3}$
Injection		DI	
Aspiration		turbocharged	
Exhaust		wastegate	

5.2 Surrogate and Mixture Definitions

Definition of the Gasoline Surrogate

Since pump gasoline usually includes a lot of different species and their share in the mixture varies a lot depending on manufacturer and batch, it is not feasible to model it in a 1:1 manner. Instead, substitute surrogates are chosen to model physical and chemical properties with the least number of components and as close to the experimental data as possible. The most basic method of doing so is mixing iso-octane (2,2,4 trimethylpentane, C_8H_{18}) with n-Heptane (C_7H_{16}), the resulting surrogate being called Primary Refer-

Table 5.6: Engine characteristics Engine F (Hydrogen) [87].

Property	Symbol	Value	Unit
Cylinder Count	n	6	
Bore	B_{E}	0.110 m	
Stroke	s_E	0.136	m
Stroke/Bore Ratio	s_E/B_{E}	1.23	
Displacement	V_h	1300	cm^3
Compression Ratio	CR	12	$\frac{\mathrm{m}^3}{\mathrm{m}^3}$
Injection		DI	
Aspiration		turbocharged	
Exhaust		wastegate	

ence Fuel (PRF). With a higher iso-octane share in PRF, the octane number of the fuel increases. For high octane numbers, this can lead to large Hydrogen/Carbon (H/C) ratios, heavily overestimating those of gasoline. To counteract this, toluene (C_7H_8O) is mixed into the surrogate, creating TRF. The shares are chosen depending on the knock sensitivity (Research Octane Number (RON) - Motor Octane Number (MON)) and the H/C ratio desired to achieve. To achieve a surrogate fuel that is DIN EN 228 [32] conform and consistent with the research of *Hann* [48], the surrogate definition in table 5.7 is chosen: 66.59 vol-% of iso-octane, 24.43 vol-% of toluene, and 8.98 vol-% of n-Heptane. It represents RON98 pump fuel without any ethanol content.

Volume and Mass Based Mixtures with Anisole and Cyclopentanone

Two 1:1 mixtures with gasoline each are created for Anisole and cyclopentanone; one on a mass fraction basis, the other referred to volume fraction. This is made to ensure comparability with measurement data: The reaction kinetics mechanism has been validated on the neat fuels and with a 50 % V/V mixture, for the engine test bench, only measurements for a 50 % m/m mixture are available. A comparison of laminar burning velocities, laminar flame thicknesses, and ignition delay times is given in figures 5.1, 5.2, 5.3, and 5.4. It can be observed that for both fuel mixtures, deviation does occur between the two definitions. Especially for the high temperature ignition delay times they can be as high as 12 %. This has to be kept in mind for the later validation of the knock model. To be closer to the reaction kinetic mechanism validation point, all theoretical explanation of the workflow are done with the 50 % V/V mixtures while the direct comparison to test bench data then is made with the mass-based mixtures.

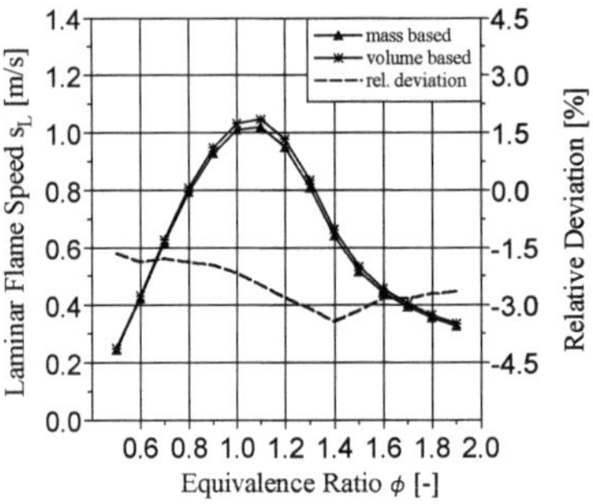

(**a**) Gasoline/Anisole

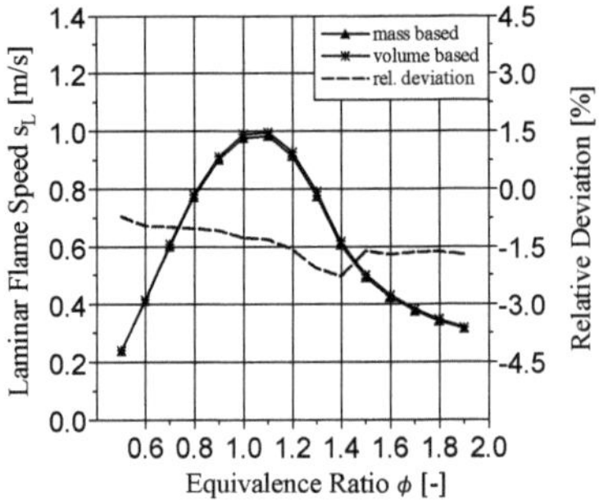

(**b**) Gasoline/Cyclopentanone

Figure 5.1: Impact of mass fraction and volume fraction based definitions on the laminar burning velocity s_L.

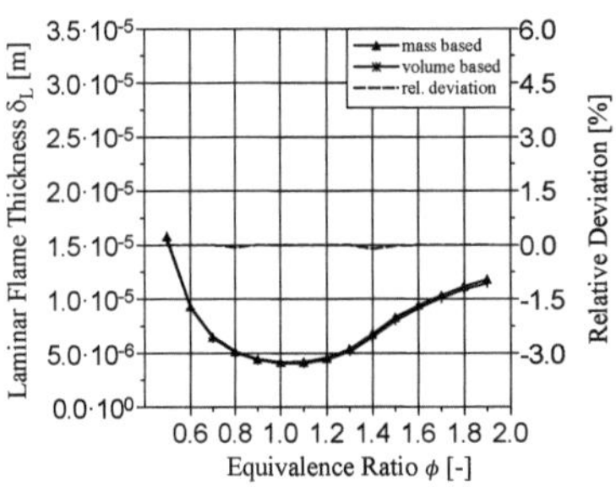

(**a**) Gasoline/Anisole

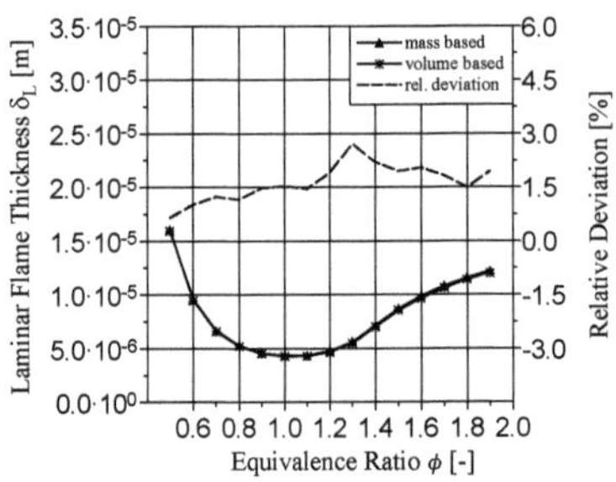

(**b**) Gasoline/Cyclopentanone

Figure 5.2: Impact of mass fraction and volume fraction based definitions on the laminar flame thickness δ_L.

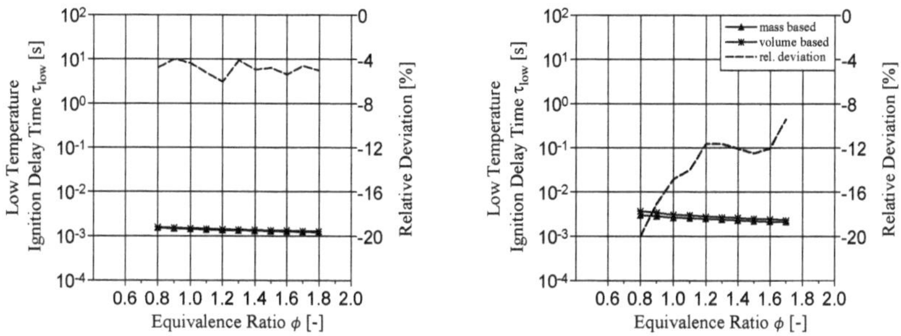

(**a**) Gasoline/Anisole (**b**) Gasoline/Cyclopentanone

Figure 5.3: Impact of mass fraction and volume fraction based definitions on the low temperature ignition delay time τ_{low}.

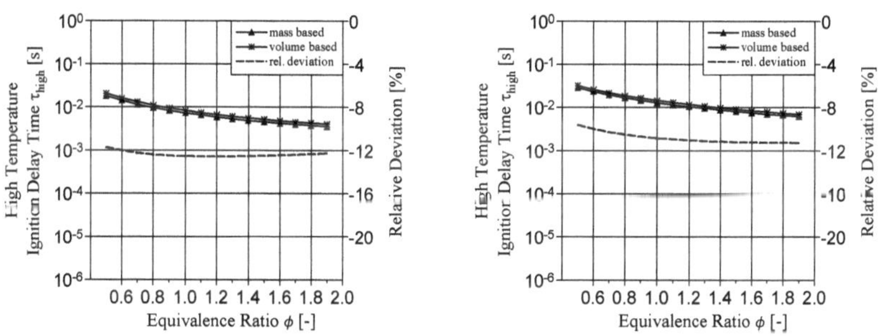

(**a**) Gasoline/Anisole (**b**) Gasoline/Cyclopentanone

Figure 5.4: Impact of mass fraction and volume fraction based definitions on the ignition delay time τ_{high}.

Mixture of Dimethyl Carbonate and Methyl Formate

For the mixture of DMC with MF, a volume ratio of 65 %/35 % is chosen. This resembles the composition given in [50] for a DMC+ mixture without any ethanol addition. Also, data is available for this mixture as it was collected within the Nachhaltige Mobilität mit Synthetischen Kraftstoffen (NaMoSyn)-project [37].

EN228 Limitation Problem for Additional Blend Components

EN228 [32] defines the minimum quality standard for European pump gasoline. Additionally to requiring fuels to have a minimum RON, it also limits the oxygen content in the fuel to 2.7 % m/m. This is done to ensure a minimum energy content for consumer protection. The problem with this reasonable limitation is the impact on possible fuel additives. As all promising candidates with high impact on the LBV and knock resistance (and thus on combustion speed and efficiency) have reasonable oxygen content, the EN228 limit is easily reached. For RON95E10, the maximum oxygen content is elevated to 3.7 % m/m. As this is already reached with the 10 % V/V of ethanol present, a further mix-in of oxygen containing additives is not possible. To still permit the creation of an EN228 "conform" fuel blend with Anisole and cyclopentanone, two possible strategies are chosen:

1. RON95 E5 is chosen as base fuel and the oxygen limits for E10 mixtures are applied. This means, starting from 2.7 % m/m oxygen content, another 1 percent point can be added by the blend-in component.
2. RON95 E10 is the basis and an arbitrary oxygen limit of 5 % m/m is applied.

Naming Conventions

To facilitate reading, naming conventions are made. From now on, fuels and fuel mixtures will be referred to as:

- TRF or Gasoline will be used as synonyms.
- E5 and E10 will be used as synonyms for TRFE5 and TRFE10
- For EN228 "conform" mixtures (maximum 3.7 % m/m of oxygen), the abbreviations $CPN_{3.70oxy}$ and $AN_{3.70oxy}$ are introduced.
- Mixtures with a maximum of 5 % m/m oxygen content are abbreviated with $CPN_{5.00oxy}$ and $AN_{5.00oxy}$.
- Other mixture names are characterized by their components, the last being followed by a fraction percentage the corresponding relation (m for % m/m, v for % V/V). An example is the blend in of 50 % m/m of Anisole

into (TRF)E10 which, according to the stated conventions, will be referred to as $E10AN_{50m}$.

Table 5.7: Mixture definitions based on volumes. Mass based definitions in table B.1, mole-fraction based definitions in table B.2.

Fuel (-mixture)	iso-Octane	Toluen	n-Heptane	Ethanol	CPN	AN
	vol-%	vol-%	vol-%	vol-%	vol-%	vol-%
"Pure" Fuels						
T(P)RF	66.59	24.43	8.98	0	0	0
T(P)RF E5	63.26	23.21	8.53	5	0	0
T(P)RF E10	59.93	21.99	8.08	10	0	0
Cyclopentanone	0	0	0	0	100	0
Anisole	0	0	0	0	0	100
50 %						
$E10CPN_{50m}$	33.7	12.36	4.54	5.62	43.77	0
$E10AN_{50m}$	34.39	12.62	4.64	5.74	0	42.61
$E10CPN_{50V}$	29.97	10.99	4.04	5.00	50	0
$E10AN_{50V}$	29.97	10.99	4.04	5.00	0	50
EN228						
$CPN_{3.70oxy}$	58.23	21.36	7.85	4.60	7.95	0
$AN_{3.70oxy}$	56.78	20.83	7.66	4.49	0	10.24
5 wt.-% Oxygen						
E10 CPN	56.28	20.65	7.59	9.39	6.09	0
E10 AN	55.03	20.19	7.42	9.18	0	8.17

6 Methodology Validation

For the validation of engine simulations, not all Measurement Data acquired on engine test benches can be used directly. Especially for combustion simulations, burn rates are of great interest as they are the immediate result of the combustion model calculations (see 2.3.1). While heat release rates can be calculated from the cylinder pressure when the crank train geometry is known, burn rates on the other hand require the use of reverse calculation (Three Pressure Analysis (TPA)) models. The data of the measured single working cycles also needs to be post-processed before it can be compared to the knock model results. In addition to that, the influence of fuel on flame wrinkling is looked at in more detail. This way, it can be judged if the existing burn rate models are still feasible even for fuels that show a high tendency to flame wrinkling (e.g., hydrogen).

For the methodology validation itself, it will be primarily done with Cyclopentanone and Anisole and their mixtures with E10 as for those fuels test bench data was best available. The main idea is to calibrate an engine model based on measurements with gasoline, then predict engine behavior when changing fuel. For this, all the data that has been collected and fit to multiple correlations (see chapters 3 and 4) is used in 0D engine simulation within the FKFS UserCylinder®.

Additionally, the methodology has been used with and validated on a few other fuels, including Methanol, a mixture of dimethyl carbonate and methyl formate, and hydrogen. Depending on the data available for those fuels, the model calibration has either been made with gasoline combustion or directly with the new fuel. Consequently, not all this data can be used for the methodology validation but in some cases it rather shows the applicability of it for additional fuel without explicitly validating the predictive capability in those cases.

S. K. Crönert, *A Complete Methodology for the Predictive Simulation of Novel, Single- and Multi-Component Fuel Combustion*, Wissenschaftliche Reihe Fahrzeugtechnik Universität Stuttgart, https://doi.org/10.1007/978-3-658-43075-7_6

6.1 TPA Model Creation

Since no TPA model exists for the current iteration of engine A, it has to be created based on the last iteration of the engine design. For this, all geometry data is reassessed and integrated into an engine model in Gamma Technologies (GT) GT-Power simulation suite.

With the utilized FKFS UserCylinder®, an iterative method to calculate burn rates from pressure traces is included. The cylinder module is fed starting values and boundary conditions (either from a 1D or 3D model and/or test bench data) and iterates the burn rate cycle per cycle until the given cylinder pressure trace is met.

As their acquisition usually is costly, no information on the internal engine component temperatures is present. The heat transfer model is thereby based on an existing calibration, adjusted to the point where the resulting burn rate after the end of combustion is nonexistent. This means there is no excessive or missing energy in the cylinder and the iterative TPA procedure does not need to adjust it by converting additional fuel. The same principle, this time applied during the compression phase, is used for determining the thermodynamic compression ratio. Starting from the geometrical ratio, this value is decreased until the TPA model doesn't need to burn reversely to decrease the cylinder enthalpy during the compression phase.

6.2 Evaluation of Knocking Operation Points

For the evaluation of the knock models for alternative fuels and the predictive capabilities of the presented methodology, data on knocking operation points for three fuels (E10, $E10AN_{50m}$, and $E10CPN_{50m}$) has been collected with the engine setup A by TME. To achieve knocking combustion, the spark advance has been increased from a stable calibration further into regions of increasing knock frequency. This has been done at two speeds, 1500 and 2000 min^{-1}, for an engine load of 15 bar. As the base compression ratio didn't lead to knocking combustion for the two fuel mixtures, it had to be increased to a geometrical ratio of 13.5. This was done by increasing the connection rod length, so the measures' influence on turbulence and load motion stayed minimal.

For each operation point, the cylinder pressure traces of 200 single cycles were measured and filtered with an 11/30 kHz bandpass filter. The Knock Peak-To-Peak (KPP) value is then extracted from this data. The knock frequency of the operation point (the chance that an average cycle at these operating conditions is knocking) is calculated by eq. 6.1. The indicator function 1_A allows for higher values of the KPP-pressure at higher engine speeds.

$$f_{\text{knock}} = \frac{\sum 1_A}{i}$$

$$\text{with:} \quad 1_A : p_{\text{kpp}}, n_{\text{eng}} \rightarrow 0, 1; \qquad \text{eq. 6.1}$$

$$1_A(p_{\text{kpp}}, n_{\text{eng}}) := \begin{cases} 1 & \text{if } p_{\text{kpp}} > \frac{n_{\text{eng}}}{1000}, \\ 0 & \text{else.} \end{cases}$$

Plotting the results of this evaluation versus their corresponding operation point - in this case, the location of ϕ_{50} - results in the figure 6.1. The threshold line at 10 % indicates the point where, if exceeded by the knock frequency, the operation point is treated as knocking. It can be seen, that apart from $\text{E10AN}_{50\text{m}}$ and $\text{E10CPN}_{50\text{m}}$ having lower knock tendencies and thus achieving a better location of the mass fraction burnt 50 %, their sensitivity to changes in the spark advance is higher. This means, once their state is close to the knock boundary, even slight advances in the ignition angle increase the knock frequency by relevant amounts. This is assumed to be based on the higher burning velocities (higher LBVs and earlier combustion with increased turbulence effects) of the 50 % m/m mixtures, resulting in lower cycle-to-cycle variations. This has to be taken into consideration when simulating optimum operation points and calibrating them for engines that are run with high burning velocity fuels.

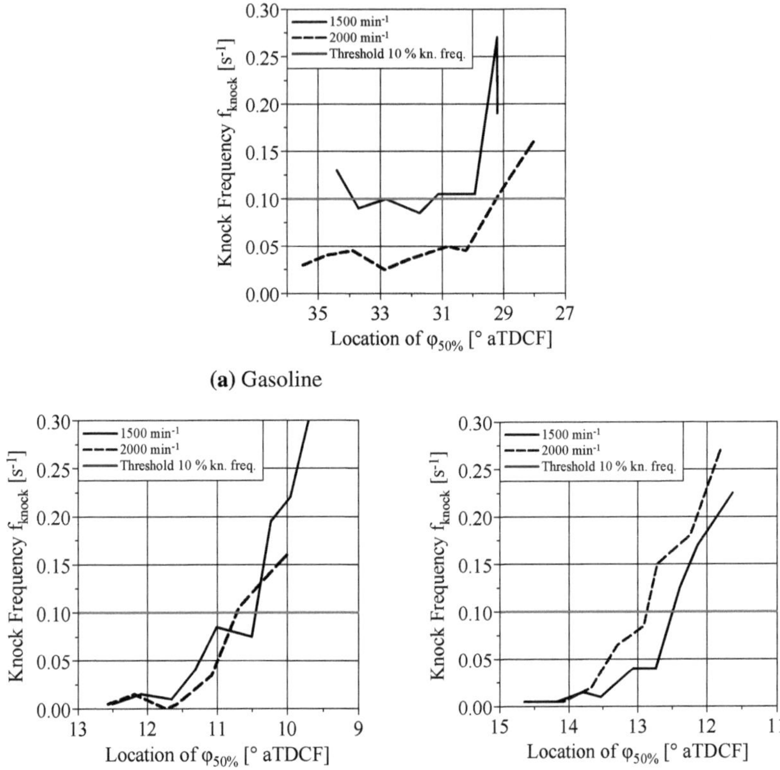

(a) Gasoline

(b) Gasoline/Anisole

(c) Gasoline/Cyclopentanone

Figure 6.1: Knock frequencies for a 15 bar, 2000 min^{-1} operating point for gasoline, $E10AN_{50m}$, and $E10CPN_{50m}$ on Engine A.

6.3 Fuel effects on Flame Wrinkling and the Early Flame Kernel

In 2.2.6 it was established, that the *Lewis* Number is a quantity qualified to determine the grade of wrinkling that occurs in the flame front. Especially for the early flame kernel development phase, when the laminar kernel slowly grows and turbulent combustion gradually becomes more dominant, it is assumed that flame wrinkling effects have an impact on how the early phase of the burn rate is formed. In an attempt to predict the fuel influence on this early phase of combustion, *Lewis* numbers of all fuels are calculated, both for low pressures and temperatures and for engine relevant boundary conditions. For the comparison between gasoline (E10) and E10CPN$_{50m}$, or E10AN$_{50m}$ respectively, these values are shown in figures 6.2 and 6.3.

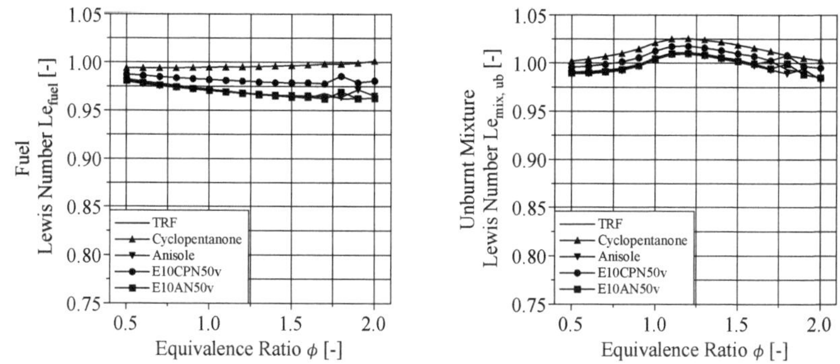

(a) Fuel at STP

(b) Unburnt mixture at STP

Figure 6.2: Calculated Lewis numbers for fuel (-mixtures) and the unburnt mixture at standard temperature and pressure. Some numerical fragments in the calculated values show at higher equivalence ratios.

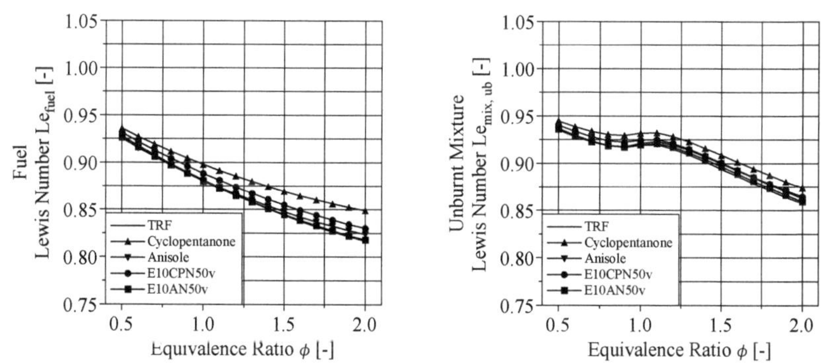

(a) Fuel at 800 K and 50 bar

(b) Mixture at 800 K and 50 bar

Figure 6.3: Calculated Lewis numbers for fuel (-mixtures) and the unburnt mixture at engine relevant operating conditions.

The calculations show that with increasing temperature and pressure, the *Lewis* number decreases, showing that thermal diffusion becomes less relevant than species diffusion. The additional consideration of unburnt air in the mixture has little effect on the absolute *Lewis* numbers, but obviously changes the sensitivity to the equivalence ratio. This would mean that for engine relevant operation conditions, flame wrinkling even increases compared to standard conditions. From the comparison between the fuel(-mixtures), one would assume, that the difference in the *Lewis* number for pure Cyclopentanone and the $E10CPN_{50m}$ mixture would be visible for the early burning phase of Cyclopentanone combustion. If this was the case, the phase of the early flame kernel should burn slower than the mixtures Cyclopentanone is compared to.

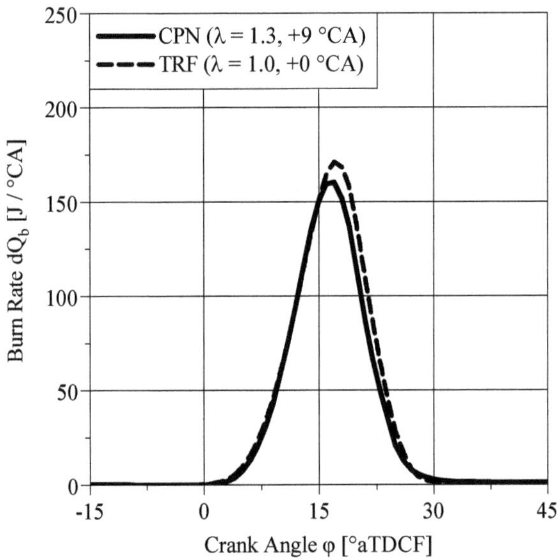

Figure 6.4: TPA results for Cyclopentanone (at $\lambda = 1.3$ and a +9 ° x-offset) and gasoline burn rates at 1500 min^{-1} and 15 bar load on Engine A.

The TPA-results for Cyclopentanone and gasoline for a high load, low speed operation point (1500 min^{-1}, 15 bar) are shown in figure 6.4. To facilitate the comparison, for Cyclopentanone, a mixture with similar burn rate maximum to the one of gasoline at stoichiometric conditions is chosen. Additionally, the curve for Cyclopentanone his offset by positive 9 °CA, so that the start of combustion lies at approximately the same time. With this setup, and assuming roughly the same effects on the *Lewis* number for pressure and temperature variations for all fuels, one is looking at a *Lewis* number of Cyclopentanone being 5 % higher than that of gasoline. Judging from this, either there is no significant fuel dependent effect on flame wrinkling during the early flame kernel development, or the flame wrinkling itself has no major influence on the first phase of combustion. One factor might be the increased turbulence effects of earlier combustion phasing that could neglect any minor influences. With Cyclopentanone and Anisole having laminar burning velocities higher than those of gasoline, this could also be mitigating any flame wrinkling effects as they are opposing each other. Although, if the latter was the case, it should be noticeable with Anisole combustion as well as it almost has the same *Lewis* number as the gasoline base fuel. Since this is not the case, at least for those two fuels any fuel influences on flame wrinkling and the early flame kernel development can be neglected. This also means that there is no need to specifically address these effects in later simulations.

6.4 Cyclopentanone and Anisole

Cyclopentanone and Anisole have been in the focus of the FVV research project *Fuel Composition for CO_2 Reduction*, as they promised high potential for efficient combustion relying on high laminar burning velocities in combination with high knock resistance. To be able to use the presented methodology with those fuels, a reaction kinetic submechanism had to be created. For this case, HGD and ITVTC Aachen gathered ignition delay time and laminar burning velocity data (the corresponding procedures are described in chapter 2).

With the created submechanisms, the toolchain for the acquisition and the correlation fit of laminar burning velocities, laminar flame thicknesses and ignition delay times can be used. The variable's data for all fits is then coded into the cylinder compound fortran code and a new binary is created. With this binary, all additional fuel data collected for submodel calibration, and the TPA results, the engine process calculation is performed. In this case, the TPA provides data for cylinder charge and mixture temperatures. This is only done for validation purposes and no limit of the calculation models. For an upcoming release, the additional option of passing correlation variables directly to the cylinder module is planned. With this, there is no need for a new, hard coded release for every additional fuel mixture.

When comparing the fundamental models described in section 2.3 and the fuel influence onto them, it can be seen, that there are basically only two major models the presented methodology needs to be validated for: The combustion model for the burn rate prediction and the ignition delay and the knock model for the prediction of the KLSA. The goal for the prediction hereby lies in matching the post processed engine test bench data. In the following, the result of the model predictions and test bench measurements will be compared to each other for an equivalence ratio variation at 15 bar load and 1500 min^{-1}.

6.4.1 Burn Rate Prediction

The base calibration of the combustion model has been done with the *Damköhler* model and TRFE10. At the 15 bar, 1500 min^{-1} operation point, the turbulence pre-factor has been scaled for the simulated burn rates to match the TPA results. For the eleanment variation for $1.0 \leq \lambda \leq 1.76$ (leanest stable operation point), this is shown in figure 6.5. It can be seen, that the burn rate curves for the simulated operation points (solid) match the ones from the test bench measurements (dashed) really well. The equivalence ratio dependence for the laminar flame speed seems to be modeled correctly, as the burn rate maximum scales well with the mixture composition. Only for operation points with medium air-to-fuel ratios, a slight deviation at the burn rate maximum is noticeable. If this again shows up for the other fuels, it might be caused by engine related (piston shape, cylinder head shape), three dimensional effects not coverable by QD models. Additionally, the early phase of combustion is under-predicted by the simulation, the combustion in the TPA results starts earlier and the conversion rate rises less quickly. It is assumed, that this is caused by overlapping single working cycles with slightly changing centers of combustion that cause this averaging effect. This is congruent with the findings of *Hann* [48] and would also explain the slightly smaller maxima in the burn rate. Overall, the calibration for the base fuel is adequate and it can be used for the prediction for fuel changes.

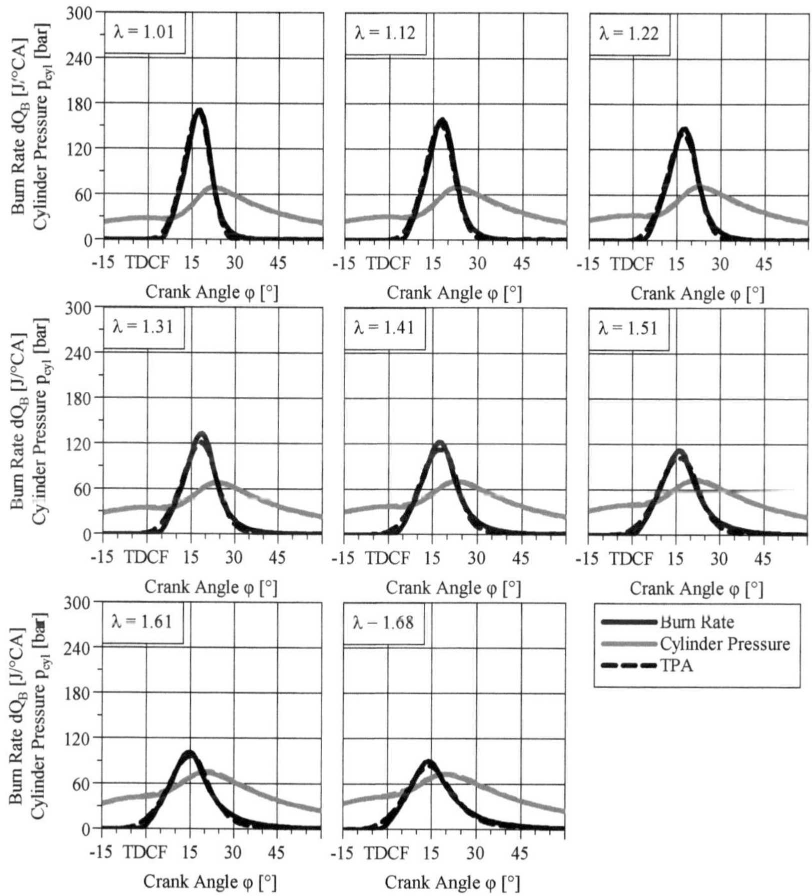

Figure 6.5: Base combustion model calibration of engine A with TRFE10. Operation point at 15 bar load and 1500 min^{-1} speed, enleanment variation.

Anisole

First, the combustion for the neat fuels is predicted, starting with Anisole. No recalibration is made, all models use the same parameters as with the TRFE10 combustion simulation. The results for Anisole combustion can then be seen in figure 6.6. It is already noticeable, that with the knock resistance of Anisole being greater than that of gasoline, the center of combustion is earlier. With this, the higher turbulence level closer to the TDC can be used. The Anisole also profits from its higher burning velocity overall, resulting in faster combustion with a higher thermodynamic efficiency. Additionally, the maximum enleanment with stable combustion has slightly shifted, with $\lambda = 1.86$ being the leanest mixture for pure Anisole that still burns reliably. This correlates to the lower flammability observations made in subsection 2.2.8. The prediction quality for pure Anisole is great, especially regarding the fact, that no recalibration has been done. The quality of the prediction is solely based in the adequate depiction of the fuel properties that has been created with the workflow described in chapter 4. Even if the burn rate maximum is encountered a bit later than with the TPA results, the congruence is perfectly sufficient for engine simulation.

Cyclopentanone

For the switch to Cyclopentanone, similar results are observed. While the deviation for medium AFR values seems more exaggerated as with TRFE10, the predictions still show great quality. Especially the high lean burn limit for pure Cyclopentanone shows how promising this fuel could be for high efficiency lean operations.

E10AN$_{50V}$

Again for the E10AN$_{50V}$ mixture, the small over-predictions at medium AFRs show up. Additionally, at the lean burn limit, the burn rate of the TPA results is slower than the simulation model predicts. As the cyclic variations at this operation close to the burn limit are already quite high ($COV \geq 10\,\%$), this is most certainly caused by higher deviations in the center of combustion. Even if the mixture and the boundary conditions are quite similar to those of the previous operation point, the TPA results change a lot. This assumption is strengthened, since the base calibration

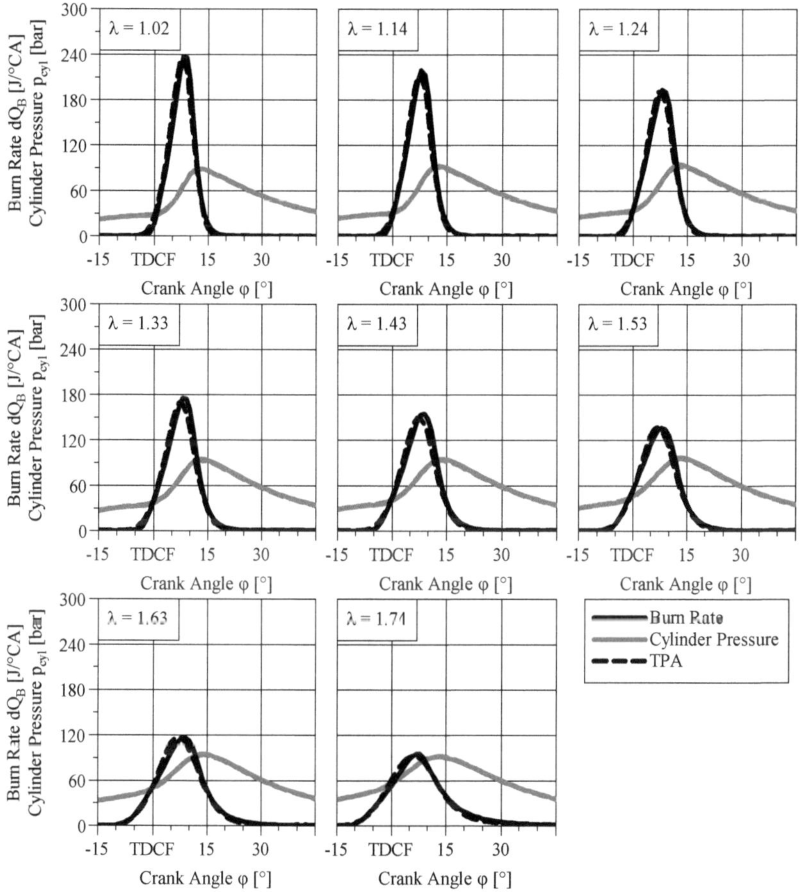

Figure 6.6: Combustion model burn rate prediction for engine A with a fuel change to Anisole. Operation point at 15 bar load and 1500 min^{-1} speed, enleanment variation. The simulation models parameters all stay the same as with TRFE10.

for the Cycle-to-Cycle Variations (CCV)-model dampens the burn rate and ends in almost the exact same shape as with the TPA results.

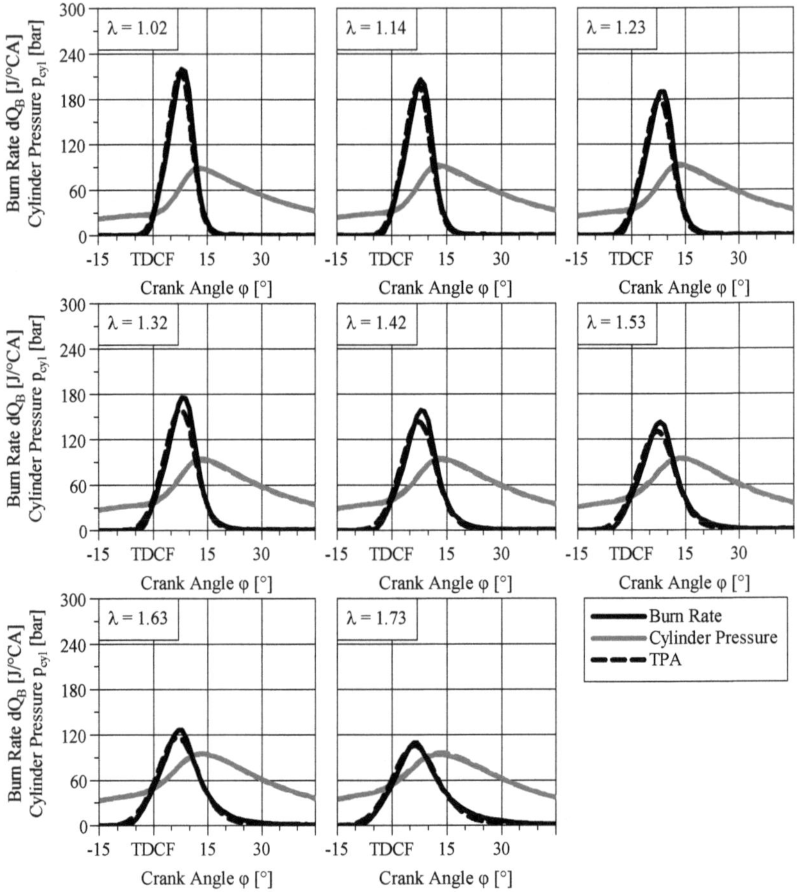

Figure 6.7: Combustion model burn rate prediction for engine A with a fuel change to Cyclopentanone. Operation point at 15 bar load and 1500 min^{-1} speed, enleanment variation. The simulation models parameters all stay the same as with TRFE10.

E10CPN$_{50V}$

The results for E10CPN$_{50V}$ are just shown for completeness, as they basically have the same features as the previously discussed neat fuels and fuel mixtures.

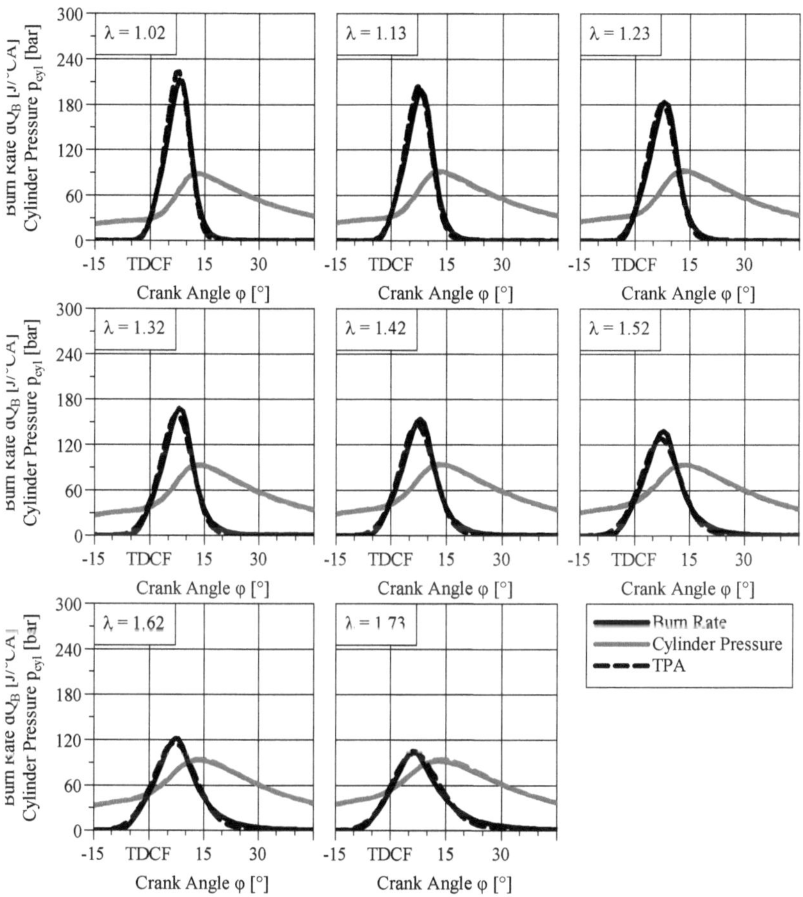

Figure 6.8: Combustion model burn rate prediction for engine A with a fuel change to $E10AN_{50V}$. Operation point at 15 bar load and 1500 min^{-1} speed, enleanment variation. The simulation models parameters all stay the same as with TRFE10.

Comparison of Burn Rate Models

All previous investigations were done based on the *Damköhler* turbulence model. To make sure the recently implemented *Peters*-model also works with the presented approach, a quick comparison is made with the two mod-

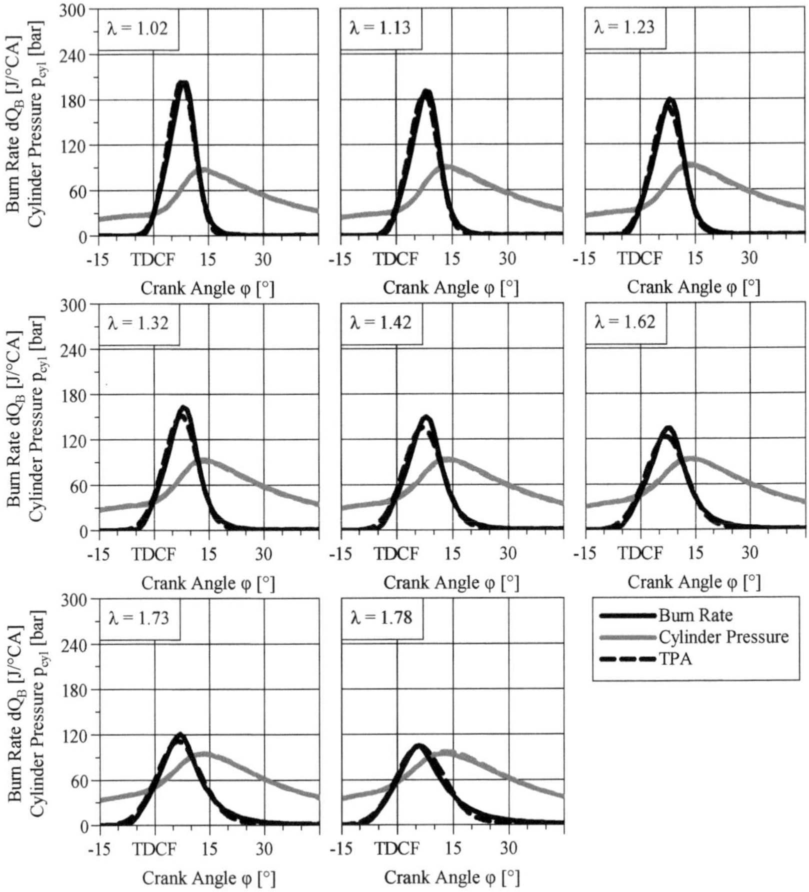

Figure 6.9: Combustion model burn rate prediction for engine A with a fuel change to $E10CPN_{50V}$. Operation point at 15 bar load and 1500 min^{-1} speed, enleanment variation. The simulation models parameters all stay the same as with TRFE10.

els and the prediction of $E10CPN_{50m}$ combustion. Figure 6.10 shows the same operation points as figure 6.9, just with an exchanged combustion model. It can be seen, that with the *Peters*-model, the trend in the burn rate maxima for the enleanment variation is better met. Also, the predicted burn rate at stoichiometric conditions basically lies directly on the one from the TPA investigation. On the other hand, this combustion model struggles to match the short and abrupt end of combustion phase for the operation points with faster combustion - this also shows for the base calibration with gasoline. Apart from that, this comparison shows that the predictive qualities of the workflow do not depend on the combustion model chosen.

Additionally, it has to be noted, that with the two combustion models usually one of them fits certain engine types better than the other. For example, the *Peters*-model shows better congruence with high displacement industrial natural gas engines while the *Damköhler*-equation yields better results for really fast combustion in high-turbulence engines with smaller displacement. The results show, that the combustion model does not matter for the predictive qualities for a fuel change, one should only stay with the same model as the calibration has been made with since the model parameters are not directly translatable.

6.4.2 Knock Occurrence Prediction

For the knock occurrence prediction, the compression ratio of the research engine had to be changed to 13.5:1, as with the base configuration (compare with table 5.1) no knocking combustion could be observed within the engines operation range. Even with this increase in the compression ratio, the pure fuels did not show any signs of knocking. For this reason, the predictive quality of the knock models was only looked at for the E10, $E10AN_{50V}$, and $E10CPN_{50V}$ fuels. As the *Hess*-model was still in development during the course of this work, primarily the *Fandakov* knock criterion has been used. Figure 6.11 shows its base calibration for a center of combustion variation with E10 at two engine speeds and a 15 bar load. The line with triangle markers shows the knock frequency measurement taken at the test bench. As (somewhat arbitrary but common) knock frequency threshold, 10 % of knocking cycles are chosen. Once an operation point exceeds this fraction (grey dashed line), it is regarded as a knocking operation point. For the calibration of the *Fandakov* knock model, the threshold value $w_{ub,bl}$ has been adjusted until the simulated KLSA roughly met the measured one. As the knock model can only tell if an average cycle is showing knocking combus-

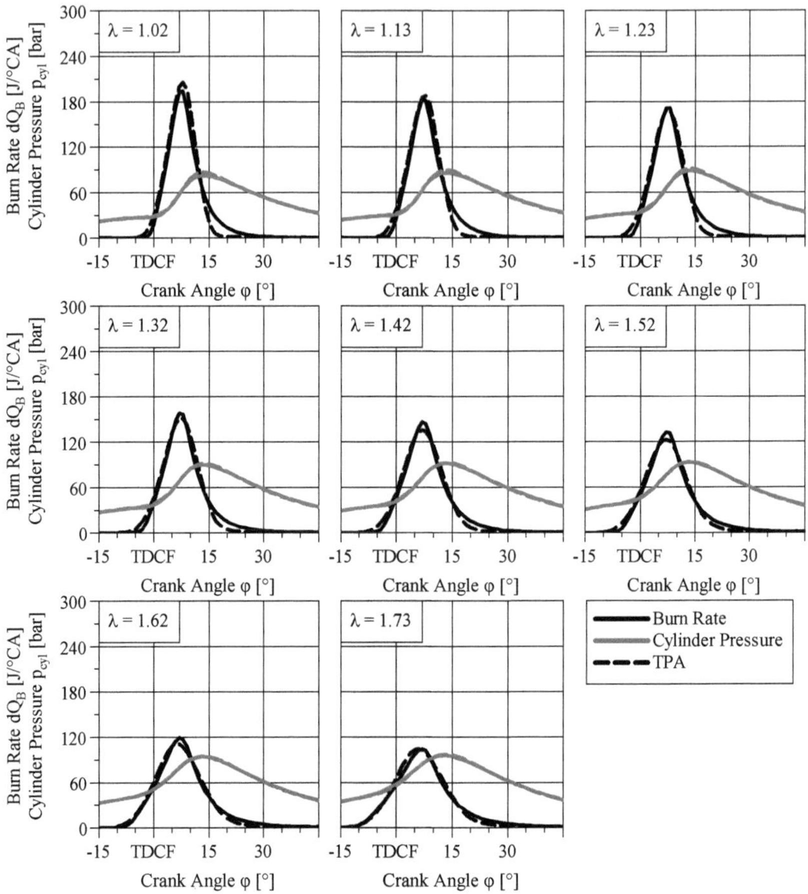

Figure 6.10: Peters Combustion model burn rate prediction for engine A with a fuel change to $E10CPN_{50m}$. Operation point at 15 bar load and 1500 min^{-1} speed, enleanment variation. The simulation models parameters all stay the same as with TRFE10.

tion (knock frequency $\geq$ 10 %), the model output is drawn in black boxes, only telling if an operation point is knocking or not (left y-axis). The base fuel shows unexpected behavior for the lower engine speed, and the model cannot meet the measurements at every calibration point. For later centers of combustion, one would expect the knock tendency to decrease but this is not the case. It is assumed that this is related to faulty measurements or changes in boundary conditions that are not recognizable in the data for the measurement points available.

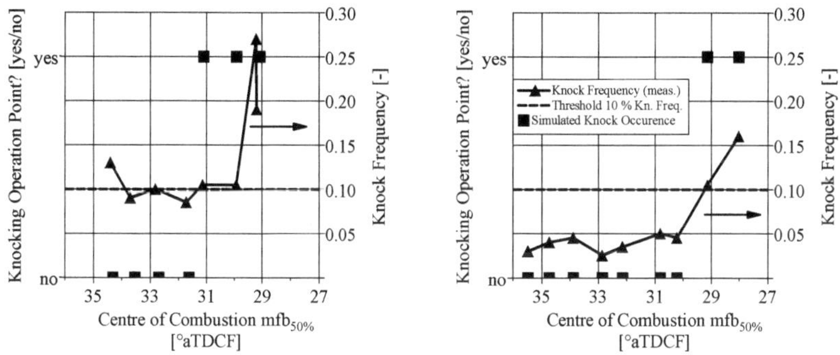

(**a**) 1500 min^{-1} (**b**) 2000 min^{-1}

Figure 6.11: Base calibration of the Fandakov knock model with E10.

With the Anisole/E10 mixture $E10AN_{50V}$, in figure 6.12 it can be seen that the knock limit is met with sufficient accuracy. With the error being within the same order of magnitude as with the base calibration, the demands in respect to the predictive quality are met. Additionally, two interesting observations can be made with the measurements for the fuel mixture: On the one hand, the center of combustion that is limited by knock occurrence has shifted from around 29 °aTDCF to 11 °aTDCF. On the other, the knock boundary became significantly thinner. While with the base fuel, the knock frequency slowly increased with the shift to earlier ϕ_{50}, with the $E10AN_{50V}$ mixture, there is less than a degree difference between the locations of a knocking operation point and an operation point that does not show any signs of knocking at all. This is most certainly based on the higher laminar burning velocities (see 3.2) reducing the CCVs. With a more stable combustion, either most of the single working cycles are knocking or not. There is no big variation between the working cycles and thus the transition becomes more sharp.

The results for the $E10CPN_{50V}$ mixture are shown in 6.13: as with the Anisole mixture, the steep gradient of the knock frequency at the knock limit is

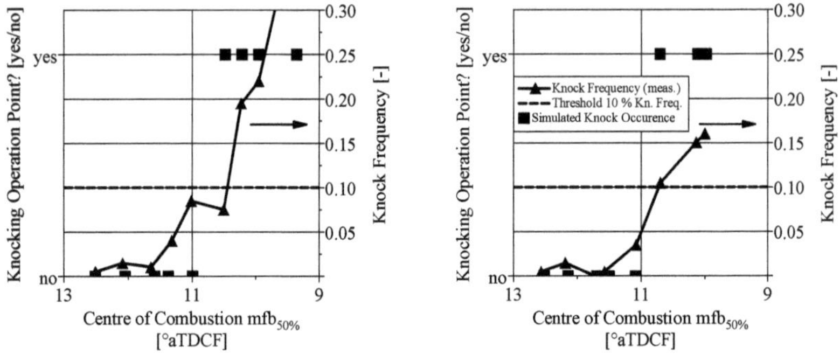

(a) 1500 min^{-1} (b) 2000 min^{-1}

Figure 6.12: Knock occurrence prediction with the Fandakov knock model and $E10AN_{50m}$.

recognizable. For the higher engine speed, the knock boundary predicted by the model starts earlier (with later ϕ_{50}), although the error is below $1\,°CA$. This means that for both fuels, the predictions of the *Fandakov* model are great for a first estimation of how a change in fuel would influence the engines resistance against knock occurring.

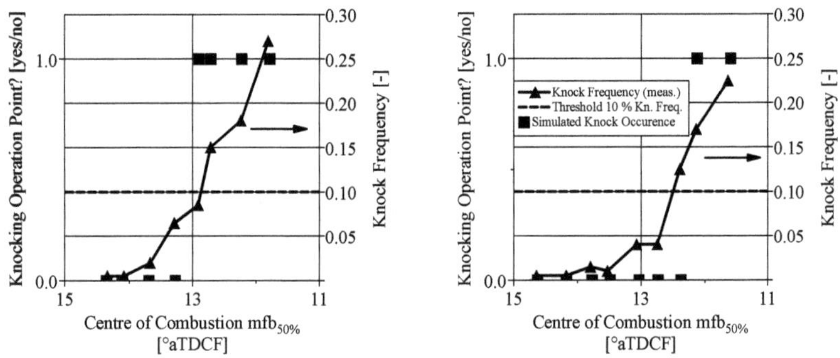

(a) 1500 min^{-1} (b) 2000 min^{-1}

Figure 6.13: Knock occurrence prediction with the Fandakov knock model and $E10CPN_{50m}$.

As the final work on the *Hess* model was finished just recently, it has only undergone a small check to judge if it is also feasible for predictions with the presented methodology. Unfortunately, no calibration could be found that matched the measured knock onset values for all three fuel(-mixtures), let alone one that could predict the results for the new fuels only with a calibration for gasoline. As the graphs in figure 6.14 show, the resulting Π for the fuel mixtures has a big offset compared to the base calibration with gasoline. This might be caused by the difference in combustion locations as the knock resistance of the mixtures is significantly higher. Nevertheless, the curves for the resulting Π values show a promising correlation, so that this could be compensated for in a future version of the knock model. In the meantime, really good results have been achieved with a really quick readjustment of the model parameters for each fuel. This obviously limits the predictive capabilities but is perfectly fine if measurement data is available. This way, for both fuels a great agreement between measurement and simulation results has been achieved.

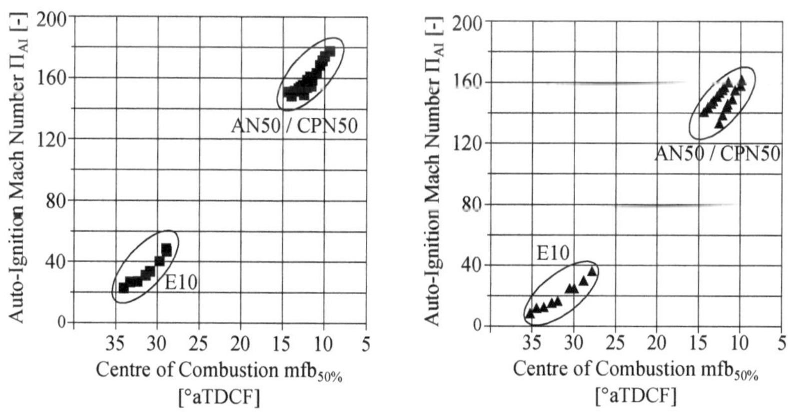

(a) 1500 min^{-1} **(b)** 2000 min^{-1}

Figure 6.14: Calculated auto-ignition mach numbers Π_{AI} for the base fuel and the fuel mixtures $E10AN_{50m}$ and $E10CPN_{50m}$ at 1500 and 2000 min^{-1}.

For all knock models used, it has to be noted that with them being extremely sensitive to engine geometry and especially the cylinder temperature, predictive simulations should always be handled with extreme caution. They might be accurate enough to give a first estimate for the knock resistance of a fuel, but should always be validated with engine measurements. With the steep slopes of the ignition delay times, even a 5 K deviation in the cylinder temperature can result in the KLSA being shifted a few degrees. With the low amount of validation data, even the good results for the *Fandakov* knock model should be treated sceptically. The experience with other research projects (e.g., [7]) showed, that the fraction of unburnt mass in the boundary layer may be unfeasible to accurately and explicitly judge if an auto-ignition occurrence results in knocking combustion or not.

6.5 Methanol

Methanol combustion has been predicted for and validated on engine B. The results for an enleanment variation are shown in figure 6.15. As the compression ratio in this variation has been quite low compared to engine A, the absolute numbers for the burn rate of methanol cannot directly be compared to those of other fuels. Though it can be seen, that the absolute values and trends of methanol combustion are well met. Even if the burn rate maximum and its location slightly deviates from the measured values, the overall quality of the predictive simulation is sufficient.

6.6 Dimethyl Carbonate/Methyl Formate Mixture

For the simulation of burn rates of 65/35 % V/V mixture of DMC with Methyl Formate ($DMC_{65}MF_{35V}$) within the *NaMoSyn* project funded by the german BMBF, the same methodology has also been applied. With it, the combustion of a single cylinder engine (engine C 5.3) and a multi cylinder engine (engine D 5.4). Their respective simulation results are shown in the figures 6.16 and 6.17. Although in this case, the combustion model was only calibrated to the burn rates with $DMC_{65}MF_{35V}$ and no predictions were made on the basis of gasoline combustion, this still shows, that the combustion of $DMC_{65}MF_{35V}$ can be matched.

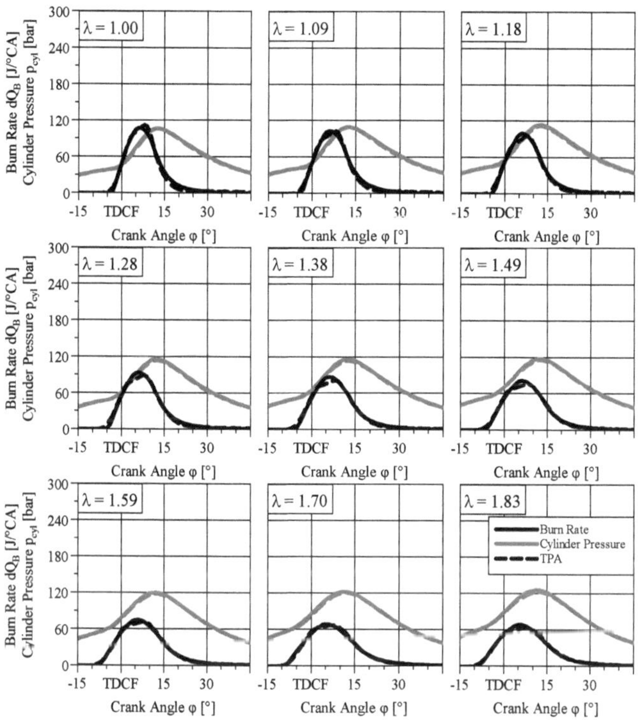

Figure 6.15: Validation of predicted methanol combustion on engine B at 15 bar load and 2500min^{-1} engine speed, enleanment variation.

6.7 Hydrogen

The integration of Hydrogen into the combustion simulation was really challenging. While the ignition delay times show standard, single domain Arrhenius behavior, the fitting of the laminar burning velocity data was more difficult than with the other fuels. Since the burning velocities are almost an order of magnitude higher than with the fuels commonly used in internal combustion engines and the trend for high dilutions is less prominent, the standard approach connected to the works of *Hann* [48] and *Mueller* [84] was not feasible any more. Thus, the set of equations had to be adjusted to better mirror the sensitivities of hydrogen to changing boundary condi-

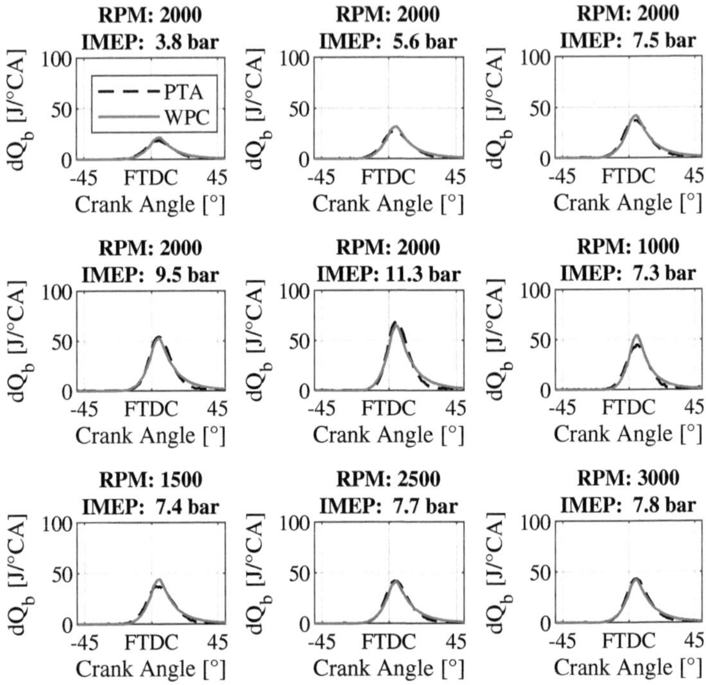

Figure 6.16: Simulated versus PTA burn rates for $DMC_{65}MF_{35V}$ compared for engine C (for specs see table 5.3)[5].

tions. As for all other fuels, the base correlation was only used for empirical fits and the results for e.g., the inner layer temperate T_0 did not necessarily correlate to physical numbers, no problem with an empirical adjustment of those equations has been identified.

With those adjusted equations integrated into the automated processes, the laminar burning velocity data of hydrogen could be fitted within reasonable error ranges. Currently, publications for two engines models created with this hydrogen model exist: in a cooperation of the FKFS and Bosch, the combustion model has been calibrated to an engine based on the MTU BR2000 and then multiple parameter variations were made to extensively demonstrate the predictive capabilities of the modeling approach. Figure 6.18 shows, that a good agreement between prediction and measurement

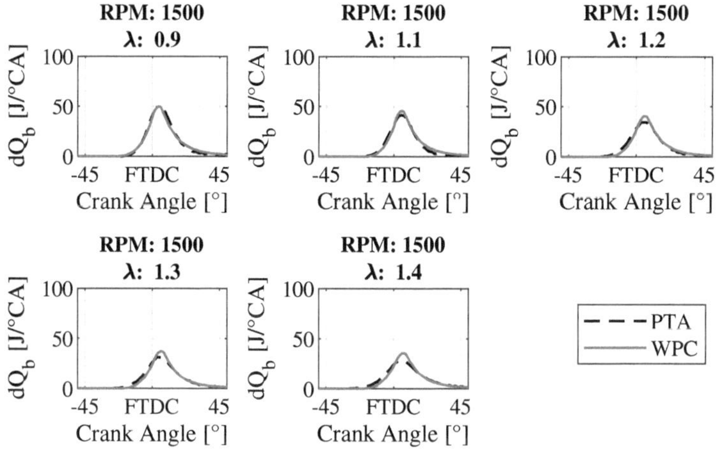

Figure 6.17: Simulated versus PTA burn rates for $DMC_{65}MF_{35V}$ compared for engine D (for specs see table 5.4)[5].

has been achieved [64]. This explicitly is true for variations of the load, the ignition timing, and the EGR ratio.

In figure 6.19, the deviation in IMEP between measurement and simulation is shown. It can be seen, that the high quality of the prediction is found for a wide range of operating points. The base offset of approximately 0.205 bar is caused by an overestimation of the fuel mass due to a lower quality of cylinder pressure data, only the deviations around this offset are due to inaccuracies in the combustion model and the presented methodology. Nevertheless it can be seen, that the achieved quality is sufficient for predictive engine simulations and the methodology is applicable for hydrogen combustion as well. The operation points correlate to:

- Operation Point 1-12: Ignition advance without EGR at low load
- OP 13-24: Ignition advance with low EGR at low load
- OP 25-36: Ignition advance with medium EGR at low load
- OP 37-50: Ignition advance with high EGR at low load

- OP 51-57: Ignition advance without EGR at high load
- OP 58-68: Ignition advance with low EGR at high load
- OP 69-79: Ignition advance with high EGR at high load

Additionally, the same procedure has been used by Deutz to simulate the combustive behaviour of engine F for a fuel change from Diesel to Hydrogen and then optimize the combustion dynamically for minimal NO_x emission during load steps by fitting a neural network to the 0D/1D simulation results [87]. Unfortunately there have been no publications of burn rate comparisons for this engine, so in this case, the prediction quality of the simulation can not be judged precisely.

6.8 Limitations of the Presented Methodology

When looking at the results for the combustion simulations in this chapter, it can be stated that over the wide range (and diversity) of fuels tested, the presented methodology shows great performance. The same applies to the engine designs and geometries. Though, there are a few remarks to be made to those who want to implement it into their workflow:

Firstly, even if a wide range of diverse fuels with big differences in their properties has been looked at, there is no guarantee that this methodology works for every fuel.
Secondly, the calculation results for the reaction kinetics simulations are highly dependent on the quality of the mechanisms used. It is thereby advised to only use mechanisms that have been explicitly created and validated for the fuel and purposed they are used. While smaller errors for the laminar burning velocity and the laminar flame thickness don't have bigger impacts on the combustion simulation, the misprediction of ignition delay data can induce bigger deviations in the simulation of knock limits.
And *lastly*, the mechanism creation relies on the measurement of laminar burning velocities and ignition delays. With the vague repeatability of those measurements and overall low precision (e.g., multiple ways to extrapolate the data to unstreched flames leading to different results), even for validated mechanisms, there is still room for deviations and error.

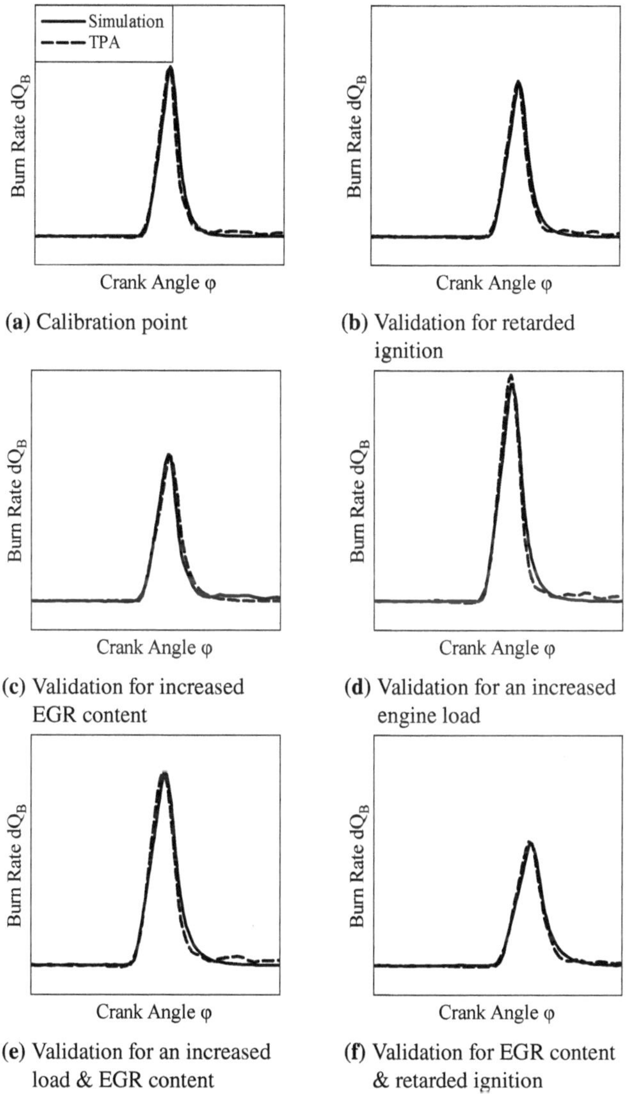

(**a**) Calibration point

(**b**) Validation for retarded ignition

(**c**) Validation for increased EGR content

(**d**) Validation for an increased engine load

(**e**) Validation for an increased load & EGR content

(**f**) Validation for EGR content & retarded ignition

Figure 6.18: Calibration and validation of the Hydrogen combustion model on engine E by the means of burn rates. Solid lines show the simulation results, dashed lines are taken from a TPA done on test bench measurements [64].

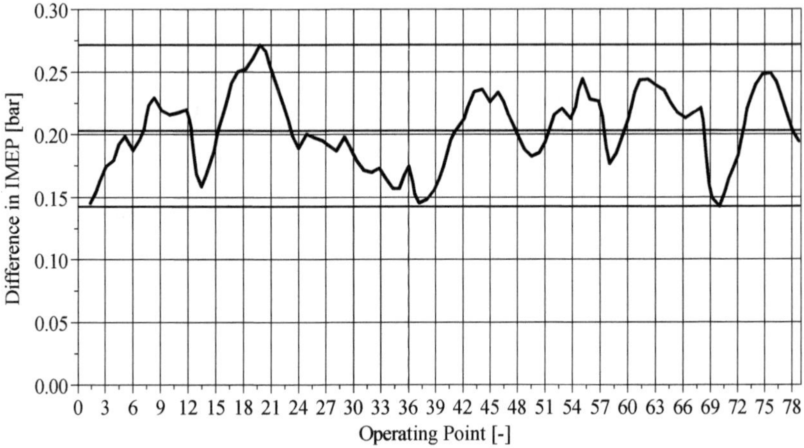

Figure 6.19: Predictive quality of the hydrogen combustion simulation over a wide range of boundary conditions. Adapted from [64].

7 Conclusion and Outlook

It has been shown, that with current QD simulation models that are based on the physical and chemical properties of fuels, it is possible to accurately predict engine combustion solely based on the fuel's characteristics. This builds on the findings of *Hann* who was able to predict fuel dependent combustion behavior based on the laminar burning velocities of fuels (mainly natural gas fuels of changing composition). For this purpose, laminar burning velocities, laminar flame thicknesses, low and high ignition delay times, and temperature increases by the first stage of ignition have been calculated with the help of distributed computing and cloud computing. All other fuel characteristics have been taken from well established chemistry databases (e.g., NIST Chemistry Webbook). With these findings, it is feasible to automate the whole calculation and correlation fitting process so that the implementation of a new fuel is dependent only on computational resources. This can be important especially for following research projects where the main focus can be shifted towards the application of novel fuels in engines rather than on their base characteristics.

The same method has also been applied to fuel mixtures - with the adapted *Fandakov* ignition delay correlation, the reaction kinetics calculations can be the same as with the pure fuels. For all other properties, mixing rules have been implemented and applied to accurately predict the behaviour of arbitrary fuel compositions in engine simulation.

The combustion behavior of all fuels and fuel mixtures could be validated against a wide variation of engines and operation points. This way it has been proven that the methodology of automating fuel property acquisition and correlation fitting is accurate enough for predictive engine simulations. Although every newly implemented fuel or fuel mixture should be thoroughly checked, it is assumed that the presented methodology works for all fuels where an accurate reaction mechanism exists. Manual work is only required for choosing the appropriate ignition delay correlation and knock module.

S. K. Crönert, *A Complete Methodology for the Predictive Simulation of Novel, Single- and Multi-Component Fuel Combustion*, Wissenschaftliche Reihe Fahrzeugtechnik Universität Stuttgart, https://doi.org/10.1007/978-3-658-43075-7_7

The great advantage of the presented work shows when the potential that comes with a fuel change is shown for an existing engine. After their acquisition and integration into the calculation software, the laminar burning velocities, laminar flame thicknesses, and ignition delay times of a newly implemented fuel can be used for predictive simulations. With a full engine model calibrated to test bench measurements with gasoline, a change to methanol has been simulated on engine B. With Methanol's high burning velocities the fuel conversion can be accelerated and thus the combustion efficiency is increased. Additionally, the high knock resistance and the high enthalpy of evaporation allowed for higher compression ratios while still keeping the most efficient locations for the center of combustion. As the whole simulation was done for stationary points, possible drawbacks as subpar cold start properties have been ignored.

The results of a full engine map simulation are shown in figure 7.1: In the regions of low end torque, an improvement of 5 to 10 % in efficiency could be made, most definitely resulting from the increased burning velocities of the alcohol. As the engine load increases, so does the efficiency advantage for the Methanol combustion. With the knock tendency of gasoline gradually becoming worse for higher cylinder pressures, the fuel properties of Methanol bring the most benefit in these regions. At the knock limited high end torque, a maximum of 26 % in efficiency increase is predicted.

For optimization purposes and designs of experiments, in the future it could be a great benefit, if a fuel mixtures combustive behaviour could be accurately predicted by knowing the composition and its neat components properties. This way, not only the engine could be optimized to make use of the fuels qualities but also the fuel composition could be tweaked directly at engine relevant operating points. While absolute accuracy should be of lesser interest in this case, at least the trends have to be met. This way, optimal mixtures can be looked at in further detail.

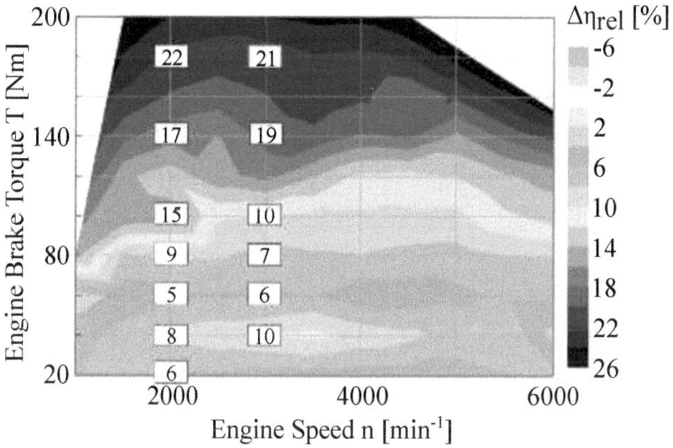

Figure 7.1: Efficiency increase with the change to Methanol in comparison to the base fuel TRF. Simulated for engine B (for specs see table 5.2). Data taken from [62].

Bibliography

[1] D. O. Allison. *Polynomial approximations of thermodynamic properties of arbitrary gas mixtures over wide pressure and density ranges*. Tech. rep. Langley Research Center, 1972.

[2] S. Arrhenius. "Über die Reaktionsgeschwindigkeit bei der Inversion von Rohrzucker durch Säuren". In: *Zeitschrift für Physikalische Chemie* 4U.1 (July 1889), pp. 226–248. DOI: 10.1515/zpch-1889-0416.

[3] R. B. Bird, W. E. Steward and E. N. Lightfoot. *Transport Phenomena*. John Wiley & Sons, 1960.

[4] N. C. Blizard and J. C. Keck. "Experimental and Theoretical Investigation of Turbulent Burning Model for Internal Combustion Engines". In: *SAE Technical Paper Series*. SAE International, Feb. 1974. DOI: 10.4271/740191.

[5] S. Blochum et al. "Potential Analysis of a DMC/MeFo Mixture in a DISI Single and Multi-Cylinder Light Vehicle Gasoline Engine". In: *SAE International Journal of Advances and Current Practices in Mobility* 4.1 (Apr. 2021), pp. 15–31. DOI: 10.4271/2021-01-0561.

[6] C. Blomberg et al. "Extension of the Phenomenological 3-Arrhenius Auto-Ignition Model for Six Surrogate Automotive Fuels". In: *SAE Technical Papers*. SAE International, May 2016. DOI: https://doi.org/10.4271/2016-01-0755.

[7] M. Blomberg et al. *Final Report to FVV Project no. 1313 Engine Knock Model*. Tech. rep. Forschungsvereinigung Verbrennungskraftmaschine (FVV), 2021.

[8] L. Boltzmann. "Studien über das Gleichgewicht der lebendigen Kraft zwischen bewegten materiellen Punkten". In: *Wissenschaftliche Abhandlungen*. Ed. by Friedrich Hasenohrl. Cambridge University Press, 1868, pp. 49–96. DOI: 10.1017/cbo9781139381420.006.

S. K. Crönert, *A Complete Methodology for the Predictive Simulation of Novel, Single- and Multi-Component Fuel Combustion*, Wissenschaftliche Reihe Fahrzeugtechnik Universität Stuttgart, https://doi.org/10.1007/978-3-658-43075-7

[9] S. Bougrine et al. “Numerical study of laminar flame properties of diluted methane-hydrogen-air flames at high pressure and temperature using detailed chemistry”. In: *International Journal of Hydrogen Energy* 36.18 (Sept. 2011), pp. 12035–12047. DOI: 10.1016/j.ijhydene.2011.06.053.

[10] F. Bozza et al. “Validation of a Fractal Combustion Model through Flame Imaging”. In: *SAE Technical Paper Series*. SAE International, Apr. 2005. DOI: 10.4271/2005-01-1120.

[11] D. Bradley et al. “Turbulent burning velocity, burned gas distribution, and associated flame surface definition”. In: *Combustion and Flame* 133.4 (June 2003), pp. 415–430. DOI: 10.1016/s0010-2180(03)00039-7.

[12] P. Burkhard et al. *Final Report to FVV Project no. 1348 Fuel Composition for CO2 Reduction*. Tech. rep. Forschungsvereinigung Verbrennungskraftmaschine (FVV), 2022.

[13] R. D. Büttgen et al. “An experimental, theoretical and kinetic modelling study on the reactivity of a lignin model compound anisole under engine-relevant conditions”. In: *Fuel* 269 (June 2020), p. 117190. DOI: 10.1016/j.fuel.2020.117190.

[14] L. Cai. “Optimized chemical mechanism for combustion of gasoline surrogate fuels”. In: *Combustion and Flame* (2014). DOI: https://doi.org/10.1016/j.combustflame.2014.11.018.

[15] L. Cai et al. “Impact of exhaust gas recirculation on ignition delay times of gasoline fuel: An experimental and modeling study”. In: *Proceedings of the Combustion Institute* 37.1 (2019), pp. 639–647. DOI: 10.1016/j.proci.2018.05.032.

[16] H. Le Chatelier and O. Boudouard. “On the flammable limits of gas mixtures”. In: *Process Safety Progress* 24.1 (2005), pp. 3–5. DOI: 10.1002/prs.10062.

[17] S. Chaudhuri, F. Wu and C. K. Law. “Scaling of turbulent flame speed for expanding flames with Markstein diffusion considerations”. In: *Physical Review E* 88.3 (Sept. 2013), p. 033005. DOI: 10.1103/physreve.88.033005.

[18] Gesellschaft Verfahrenstechnik und Chemieingenieurwesen. *VDI-Wärmeatlas*. Springer Vieweg, 2013.

[19] Z. Chen, M. P. Burke and Y. Ju. "Effects of Lewis number and ignition energy on the determination of laminar flame speed using propagating spherical flames". In: *Proceedings of the Combustion Institute* 32.1 (2009), pp. 1253–1260. DOI: 10.1016/j.proci.2008.05.060.

[20] P. Clavin. "Dynamic behavior of premixed flame fronts in laminar and turbulent flows". In: *Progress in Energy and Combustion Science* 11.1 (Jan. 1985), pp. 1–59. DOI: 10.1016/0360-1285(85)90012-7.

[21] G. B. Cook et al. "Bimolecular routes to cubic autocatalysis". In: *The Journal of Physical Chemistry* 93.7 (1989), pp. 2749–2755. DOI: 10.1021/j100344a012.

[22] S. Crönert. "Modellierung zukünftiger Kraftstoffe für die Arbeitsprozessrechnung". MA thesis. University of Stuttgart, 2019.

[23] G. Damköhler. "The effect of turbulence on the flame velocity in gas mixtures". In: *National advisory committee for aeronautics*. 1947.

[24] G. Darrieus. "Propagation d'un front de flamme". In: *La Technique Moderne and Congrés de Méchanique Appliquée Paris* (1938).

[25] D. F. Davidson and R. K. Hanson. "Interpreting shock tube ignition data". In: *International Journal of Chemical Kinetics* 36.9 (2004). DOI: https://doi.org/10.1002/kin.20024. URL: https://onlinelibrary.wiley.com/doi/abs/10.1002/kin.20024.

[26] S. Demesoukas et al. "Zero-Dimensional Spark Ignition Combustion Modeling - A Comparison of Different Approaches". In: *SAE Technical Paper Series*. SAE International, Sept. 2013. DOI: 10.4271/2013-24-0022.

[27] P. Dirrenberger et al. "Laminar burning velocity of gasolines with addition of ethanol". In: *Fuel* 115 (Jan. 2014), pp. 162–169. DOI: 10.1016/j.fuel.2013.07.015.

[28] P. Dirrenberger et al. "Measurements of Laminar Flame Velocity for Components of Natural Gas". In: *Energy & Fuels* 25.9 (Aug. 2011), pp. 3875–3884. DOI: 10.1021/ef200707h.

[29] E. Doering, H. Schedwill and M. Dehli. *Fundamentals of technical thermodynamics. Textbook for students of engineering sciences. 6. rev. and enl. ed.* July 2008.

[30] S. Dooley et al. “An experimental and kinetic modeling study of methyl formate low-pressure flames”. In: *Combustion and Flame* 158.4 (Apr. 2011), pp. 732–741. DOI: 10.1016/j.combustflame.2010.11.003.

[31] S. Dooley et al. “Methyl formate oxidation: Speciation data, laminar burning velocities, ignition delay times, and a validated chemical kinetic model”. In: *International Journal of Chemical Kinetics* 42.9 (July 2010), pp. 527–549. DOI: 10.1002/kin.20512.

[32] *EN228: Kraftstoffe für Kraftfahrzeuge - Unverbleite Ottokraftstoffe - Anforderungen und Prüfverfahren*. Technical Standard. European Union.

[33] J. Ewald. “A level set based flamelet model for the prediction of combustion in homogeneous charge and direct injection spark ignition engines”. PhD thesis. RWTH Aachen, 2006.

[34] A. Fandakov. *A Phenomenological Knock Model for the Development of Future Engine Concepts*. Springer Fachmedien Wiesbaden, Jan. 2019.

[35] A. Fandakov et al. “A New Model for Predicting the Knock Boundary with EGR at Full Load (Ein neues Modell zur Vorhersage der Klopfgrenze bei Volllast-AGR)”. In: *Ladungswechsel im Verbrennungsmotor 2017*. Springer Fachmedien Wiesbaden, 2018, pp. 47–72. DOI: 10.1007/978-3-658-22671-8_3.

[36] A. Fandakov et al. “Two-Stage Ignition Occurrence in the End Gas and Modeling Its Influence on Engine Knock”. In: *SAE International Journal of Engines* 10.4 (Sept. 2017), pp. 2109–2128. DOI: 10.4271/2017-24-0001.

[37] *Forschungsprojekt Nachhaltige Mobilität mit synthetischen Kraftstoffen*. URL: http://namosyn.de/ (visited on 17/04/2022).

[38] M. L. Frankel and G. I. Sivashinsky. “On Effects Due To Thermal Expansion and Lewis Number in Spherical Flame Propagation”. In: *Combustion Science and Technology* 31.3-4 (Apr. 1983), pp. 131–138. DOI: 10.1080/00102208308923635.

[39] I. Glassman, R. A. Yetter and N. G. Glumac. *Combustion*. Elsevier Science Publishing Co Inc, Nov. 2014. ISBN: 012407913X. URL: https://www.ebook.de/de/product/22639253/irvin_princeton_university_nj_usa_glassman_richard_a_penn_state_university_university_park_usa_yetter_nick_g_professor_department_of_mechanical_science_engineering_university_of_illinois_urbana_champaign_usa_glumac_combustion.html.

[40] S. S. Goldsborough et al. "Advances in rapid compression machine studies of low- and intermediate-temperature autoignition phenomena". In: *Progress in Energy and Combustion Science* 63 (2017), pp. 1–78. ISSN: 0360-1285. DOI: https://doi.org/10.1016/j.pecs.2017.05.002. URL: https://www.sciencedirect.com/science/article/pii/S0360128516300776.

[41] David G Goodwin et al. *Cantera: An Object-oriented Software Toolkit for Chemical Kinetics, Thermodynamics, and Transport Processes*. 2021. DOI: 10.5281/ZENODO.4527812.

[42] J. Göttgens, F. Mauss and N. Peters. "Analytic approximations of burning velocities and flame thicknesses of lean hydrogen, methane, ethylene, ethane, acetylene, and propane flames". In: *Symposium (International) on Combustion* 24.1 (Jan. 1992), pp. 129–135. DOI: 10.1016/s0082-0784(06)80020-2.

[43] M. Grill and M. Bargende. "The Development of an Highly Modular Designed Zero-Dimensional Engine Process Calculation Code". In: *SAE International Journal of Engines* 3.1 (Apr. 2010), pp. 1–11. DOI: 10.4271/2010-01-0149.

[44] M. Grill et al. "Calculating the thermodynamic properties of burnt gas and vapor fuel for user-defined fuels". In: *MTZ worldwide* 68.5 (May 2007), pp. 30–35. DOI: 10.1007/bf03226830.

[45] Ö. L. Gülder. "Correlations of Laminar Combustion Data for Alternative S.I. Engine Fuels". In: *SAE Technical Paper Series*. SAE International, Aug. 1984. DOI: 10.4271/841000.

[46] K. Stephan H. D. Baehr. *Wärme- und Stoffübertragung*. Springer-Verlag Berlin Heidelberg, 2013.

[47] W. Han and Z. Chen. "Effects of Soret diffusion on spherical flame initiation and propagation". In: *International Journal of Heat and Mass Transfer* 82 (Mar. 2015), pp. 309–315. DOI: 10.1016/j.ijheatmasstransfer.2014.11.062.

[48] S. Hann. *A Quasi-Dimensional SI Burn Rate Model for Carbon-Neutral Fuels*. Springer Fachmedien Wiesbaden, Feb. 2021. ISBN: 978-3-658-33231-0.

[49] S. Hann. "Reaktionskinetische Bestimmung laminarer Flammengeschwindigkeiten von binären, methanbasierten CNG-Substituten". MA thesis. HWTG Konstanz, 2016.

[50] M. Härtl et al. "DMC+ als partikelfreier und potenziell nachhaltiger Kraftstoff für DI Ottomotoren". In: *Zukünftige Kraftstoffe*. Springer Berlin Heidelberg, 2019, pp. 758–782. DOI: 10.1007/978-3-662-58006-6_29.

[51] R. A. Heidemann and A. M. Khalil. "The calculation of critical points". In: *AIChE Journal* 26.5 (Sept. 1980), pp. 769–779. DOI: 10.1002/aic.690260510.

[52] W. Heidemann. *Technische Thermodynamik*. John Wiley & Sons, 2016.

[53] M. Hess et al. "Knock Model Covering Thermodynamic and Chemical Influences on the Two-Stage Auto-Ignition of Gasoline Fuels". In: *SAE Technical Paper Series*. SAE International, Apr. 2021. DOI: 10.4271/2021-01-0381.

[54] M. Hess et al. "Two-Stage 0D/1D Knock Model to Predict the Knock Boundary of SI Engines". In: *Proceedings*. Springer Fachmedien Wiesbaden, 2021, pp. 514–530. DOI: 10.1007/978-3-658-33466-6_37.

[55] J. B. Heywood. *Internal Combustion Engine Fundamentals*. Ed. by A. Duffy and J. M. Morriss. McGraw-Hill, 1988.

[56] P. G. Hill and J. Hung. "Laminar Burning Velocities of Stoichiometric Mixtures of Methane with Propane and Ethane Additives". In: *Combustion Science and Technology* 60.1-3 (July 1988), pp. 7–30. DOI: 10.1080/00102208808923973.

[57] J. O. Hirschfelder, C. F. Curtiss and R. B. Bird. *Molecular Theory of Gases and Liquids*. Wiley & Sons, 1964.

[58] J. Hustad. "Experimental studies of lower flammability limits of gases and mixtures of gases at elevated temperatures". In: *Combustion and Flame* 71.3 (Mar. 1988), pp. 283–294. DOI: 10.1016/0010-2180(88)90064-8.

[59] S. Jerzembeck et al. "Laminar burning velocities at high pressure for primary reference fuels and gasoline: Experimental and numerical investigation". In: *Combustion and Flame* 156.2 (Feb. 2009), pp. 292–301. DOI: 10.1016/j.combustflame.2008.11.009.

[60] F. Joos. *Technische Verbrennung*. Springer, 2006.

[61] R. J. Kee, M. E. Coltrin and P. Glarborg. *Chemically Reacting Flow*. John Wiley & Sons, Inc., Feb. 2003. DOI: 10.1002/0471461296.

[62] V. Kelich et al. *Final Report to FVV Project no. 1307 ICE2025+: Ultimate System Efficiency*. Tech. rep. Forschungsvereinigung Verbrennungskraftmaschine (FVV), 2021.

[63] A. P. Kelley, J. K. Bechtold and C. K. Law. "Premixed flame propagation in a confining vessel with weak pressure rise". In: *Journal of Fluid Mechanics* 691 (Dec. 2011), pp. 26–51. DOI: 10.1017/jfm.2011.439.

[64] M. T. Keskin et al. *UserCylinder®: Wasserstoff-Brennverfahren*. Feb. 2022. URL: https://www.fkfs.de/fileadmin/FKFS/2_Kompetenzen/2.2_Virtuelle_Entwicklung/2.2.3_Virtuelle_Motorenentwicklung/Simulation_H2_Brennverfahren.pdf.

[65] E. Keszei. *Reaction Kinetics*. Springer International Publishing, May 2021.

[66] W. Kleinschmidt. "Klopfregelung für Ottomotoren II Trends fuür Serienentwickler". In: ed. by M. Fischer. Expert-Verlag, 2006. Chap. Selbstzündung im Klopfgrenzbereich von Serienmotoren, pp. 1–21.

[67] P. Kurzweil. *Chemie : Grundlagen, Aufbauwissen, Anwendungen und Experimente*. Vieweg + Teubner, 2012.

[68] Y. Lafay et al. "Thermal Structure of Laminar Methane/Air Flames: Influence of H2 Enrichment and Reactants Preheating". In: *Combustion Science and Technology* 181.9 (Sept. 2009), pp. 1145–1163. DOI: 10.1080/00102200902973216.

[69] K. J. Laidler. "A glossary of terms used in chemical kinetics, including reaction dynamics (IUPAC Recommendations 1996)". In: *Pure and Applied Chemistry* 68.1 (Jan. 1996), pp. 149–192. DOI: 10.1351/pac199668010149.

[70] L. Landau. "On the Theory of Slow Combustion". In: *Dynamics of Curved Fronts*. Elsevier, 1988, pp. 403–411. DOI: 10.1016/b978-0-08-092523-3.50044-7.

[71] C. K. Law and C. J. Sung. "Structure, aerodynamics, and geometry of premixed flamelets". In: *Progress in Energy and Combustion Science* 26.4-6 (Aug. 2000), pp. 459–505. DOI: 10.1016/s0360-1285(00)00018-6.

[72] Byung Ik Lee and Michael G. Kesler. "A generalized thermodynamic correlation based on three-parameter corresponding states". In: *AIChE Journal* 21.3 (May 1975), pp. 510–527. DOI: 10.1002/aic.690210313.

[73] F. vom Lehn et al. "Combustion Kinetics of the Synthetic E-Fuels Methyl Formate and Dimethyl Carbonate: A Modeling and Experimental Study". In: *30. Deutscher Flammentag, September 28-29, Hannover (Germany)*. 2021.

[74] L. Li and E. Kiran. "Estimation Of Critical Properties of Binary Mixtures Using Group Contribution Methods". In: *Chemical Engineering Communications* 94.1 (Jan. 1990), pp. 131–141. DOI: 10.1080/00986449008911460.

[75] W. Liang, F. Wu and C. K. Law. "Extrapolation of laminar flame speeds from stretched flames: Role of finite flame thickness". In: *Proceedings of the Combustion Institute* 36.1 (2017), pp. 1137–1143. DOI: 10.1016/j.proci.2016.08.074.

[76] A. N. Lipatnikov and J. Chomiak. "Turbulent flame speed and thickness: phenomenology, evaluation, and application in multidimensional simulations". In: *Progress in Energy and Combustion Science* 28.1 (Jan. 2002), pp. 1–74. DOI: 10.1016/s0360-1285(01)00007-7.

[77] C.-C. Liu et al. "High-pressure burning velocities measurements for centrally-ignited premixed methane/air flames interacting with intense near-isotropic turbulence at constant Reynolds numbers". In: *Combustion and Flame* 159.8 (Aug. 2012), pp. 2608–2619. DOI: 10.1016/j.combustflame.2012.04.006.

[78] J. C. Livengood and P. C. Wu. "Correlation of autoignition phenomena in internal combustion engines and rapid compression machines". In: *Symposium (International) on Combustion* 5.1 (1955), pp. 347–356. DOI: https://doi.org/10.1016/S0082-0784(55)80047-1.

[79] G. H. Markstein. "Experimental and Theoretical Studies of Flame-Front Stability". In: *Journal of the Aeronautical Sciences* 18.3 (Mar. 1951), pp. 199–209. DOI: 10.2514/8.1900.

[80] G. H. Markstein. "Theory of Flame Propagation". In: *Nonsteady Flame Propagation*. Elsevier, 1964, pp. 5–14. DOI: 10.1016/b978-1-4831-9659-6.50005-6.

[81] M. Matalon. "The Darrieus–Landau instability of premixed flames". In: *Fluid Dynamics Research* 50.5 (Aug. 2018), p. 051412. DOI: 10.1088/1873-7005/aab510.

[82] A. McFarlane. *Functions with Timeouts and Multiprocessing in Python*. Aug. 2020. URL: https://flipdazed.github.io/blog/quant%20dev/parallel-functions-with-timeouts.

[83] M. R. Moldover and J. S. Gallagher. "Critical points of mixtures: An analogy with pure fluids". In: *AIChE Journal* 24.2 (Mar. 1978), pp. 267–278. DOI: 10.1002/aic.690240216.

[84] U. C. Müller, M. Bollig and N. Peters. "Approximations for burning velocities and markstein numbers for lean hydrocarbon and methanol flames". In: *Combustion and Flame* 108.3 (Feb. 1997), pp. 349–356. DOI: 10.1016/s0010-2180(96)00110-1.

[85] S. P. R. Muppala et al. "Development of an algebraic reaction rate closure for the numerical calculation of turbulent premixed methane, ethylene, and propane/air flames for pressures up to 1.0 MPa". In: *Combustion and Flame* 140.4 (Mar. 2005), pp. 257–266. DOI: 10.1016/j.combustflame.2004.11.005.

[86] M. T. Nguyen, L. J. Jiang and S. Shy. "Fuel Similarity and Turbulent Burning Velocities of Stoichiometric Iso-octane, Lean Hydrogen, and Lean Propane at High Pressure". In: *26th International Colloquium on the Dynamics of Explosions and Reactive Systems*. 2017.

[87] B. Nork, A. Qriqra and R. Kleuser. "Deutz Hydrogen-ICE - a simulation of the entire system". In: *18th Symposium "Sustainable Mobility, Transport and Power Generation"*. Institute of Internal Combustion Engines and Thermodynamics (IVT), Graz University of Technology, Nov. 2021. DOI: 10.1007/s41321-021-0449-8.

[88] E. C. Okafor, Y. Nagano and T. Kitagawa. "Effects of Hydrogen Concentration on Stoichiometric H2/CH4/Air Premixed Turbulent Flames". In: *SAE Technical Paper Series*. SAE International, Oct. 2013. DOI: 10.4271/2013-01-2563.

[89] T. Wetzel P. von Böckh. *Wärmeübertragung*. Springer Vieweg, 2015.

[90] J. Pan et al. "A predictive Livengood–Wu correlation for two-stage ignition". In: *International Journal of Engine Research* 17.8 (July 2016), pp. 825–835. DOI: 10.1177/1468087415619516.

[91] Karen Schou Pedersen et al. "Viscosity of crude oils". In: *Chemical Engineering Science* 39.6 (1984), pp. 1011–1016. DOI: 10.1016/0009-2509(84)87009-8.

[92] N. Peters. *Fifteen Lectures on Laminar and Turbulent Combustion*. 1995.

[93] N. Peters. *Technische Verbrennung I*. Institut für Technische Verbrennung RWTH Aachen, 2011.

[94] N. Peters. "The kinetic foundation of thermal flame theory". In: *Internal Conference on Combustion*. 1997.

[95] N. Peters. "The turbulent burning velocity for large-scale and small-scale turbulence". In: *Journal of Fluid Mechanics* 384 (Apr. 1999), pp. 107–132. DOI: 10.1017/s0022112098004212.

[96] N. Peters. *Turbulent Combustion*. Cambridge University Press, Aug. 2000. DOI: 10.1017/cbo9780511612701.

[97] F. (Editor) Pischinger. *Sonderforschungsbericht 224*. Tech. rep. Jan. 2001. URL: http://www.sfb224.rwth-aachen.de/bericht.htm.

[98] R. Pischinger, M. Klell and T. Sams. *Thermodynamik der Verbrennungskraftmaschine*. Springer-Verlag, 2009.

[99] T. Poinsot. *Theoretical and numerical combustion*. Philadelphia: Edwards, 2005.

[100] A. Ramalingam et al. "An RCM experimental and modeling study on CH4 and CH4/C2H6 oxidation at pressures up to 160bar". In: *Fuel* 206 (2017), pp. 325–333. ISSN: 0016-2361. DOI: https://doi.org/10.1016/j.fuel.2017.06.005. URL: https://www.sciencedirect.com/science/article/pii/S0016236117306968.

[101] S. Richard et al. "On the Reduction of a 3D CFD Combustion Model to Build a Physical 0D Model for Simulating Heat Release, Knock and Pollutants in SI Engines". In: *Oil & Gas Science and Technology - Revue de l'IFP* 64.3 (May 2009), pp. 223–242. DOI: 10.2516/ogst/2008055.

[102] J. R. Rowley. "Flammability Limits, Flash Points, and Their Consanguinity: Critical Analysis, Experimental Exploration, and Prediction". PhD thesis. Brigham Young University, 2010.

[103] *San Diego Combustion Mechanism*. University of California at San Diego Mechanical & Aerospace Engineering (Combustion Research Department). Dec. 2021. URL: http://web.eng.ucsd.edu/mae/groups/combustion/mechanism.html (visited on 17/04/2022).

[104] M. Sarathy et al. "Knock Prediction Using a Simple Model for Ignition Delay". In: *SAE 2016 World Congress and Exhibition*. SAE International, Apr. 2016. DOI: https://doi.org/10.4271/2016-01-0702. URL: https://doi.org/10.4271/2016-01-0702.

[105] G. S. Settles. *Schlieren and Shadowgraph Techniques*. Springer Berlin Heidelberg, Dec. 2012.

[106] G. I. Sivashinsky. "Diffusional-Thermal Theory of Cellular Flames". In: *Combustion Science and Technology* 15.3-4 (Jan. 1977), pp. 137–145. DOI: 10.1080/00102207708946779.

[107] G. P. Smith et al. *GRI-Mech 3.0*. URL: http://www.me.berkeley.edu/gri_mech/ (visited on 17/04/2022).

[108] B. A. Stradi et al. "Reliable computation of mixture critical points". In: *AIChE Journal* 47.1 (Jan. 2001), pp. 212–221. DOI: 10.1002/aic.690470121.

[109] C.-J. Sung and H. J. Curran. "Using rapid compression machines for chemical kinetics studies". In: *Progress in Energy and Combustion Science* 44 (2014), pp. 1–18. DOI: https://doi.org/10.1016/j.pecs.2014.04.001.

[110] S. R. Turns. *An Introduction to Combustion - Concepts and Applications*. Ed. by J. John. McGraw-Hill, Inc., 1996.

[111] S. K. Upadhyay. *Chemical Kinetics and Reaction Dynamics*. Springer, 2006.

[112] J. Vancoillie et al. "A laminar burning velocity and flame thickness correlation for ethanol–air mixtures valid at spark-ignition engine conditions". In: *Fuel* 102 (Dec. 2012), pp. 460–469. DOI: 10.1016/j.fuel.2012.05.022.

[113] J. Warnatz, U. Maas and R. W. Dibble. *Combustion*. Springer Verlag Berlin Heidelberg, 2006.

[114] S. Welscher et al. *A Detailed Reaction Kinetics-Based Calculation Tool for Internal Combustion Engine-Related Ignition Processes*. Tech. rep. 2022-01-5050. SAE Technical Paper, 2022. DOI: 10.4271/2022-01-5050.

[115] C. R. Wilke. "A Viscosity Equation for Gas Mixtures". In: *The Journal of Chemical Physics* 18.4 (Apr. 1950), pp. 517–519. DOI: 10.1063/1.1747673.

[116] L. Wißmann et al. "Detailed Numerical Investigation of the Influences of External Water, Lubricating Oil and Reactive EGR Components on ICE Ignition Processes". In: *10th COMMODIA*. Engine System Division. Sapporo, Japan: Japan Society of Mechanical Engineers, July 2022.

[117] C. K. Wu and C. K. Law. "On the determination of laminar flame speeds from stretched flames". In: *Symposium (International) on Combustion* 20.1 (Jan. 1985), pp. 1941–1949. DOI: 10.1016/s0082-0784(85)80693-7.

[118] S. Yang et al. "Role of Darrieus–Landau instability in propagation of expanding turbulent flames". In: *Journal of Fluid Mechanics* 850 (July 2018), pp. 784–802. DOI: 10.1017/jfm.2018.426.

[119] K. Zhang et al. "An experimental, theoretical, and modeling study of the ignition behavior of cyclopentanone". In: *Proceedings of the Combustion Institute* 37.1 (2019), pp. 657–665. DOI: 10.1016/j.proci.2018.06.097.

[120] K. Zhang et al. "Experimental and Kinetic Modeling Study of Laminar Burning Velocities of Cyclopentanone and Its Binary Mixtures with Ethanol and n-Propanol". In: *Energy & Fuels* 34.9 (July 2020), pp. 11408–11416. DOI: 10.1021/acs.energyfuels.0c01565.

[121] V. L. Zimont. "Gas premixed combustion at high turbulence. Turbulent flame closure combustion model". In: *Experimental Thermal and Fluid Science* 21.1-3 (Mar. 2000), pp. 179–186. DOI: 10.1016/s0894-1777(99)00069-2.

Appendices

Appendix A: Nasa Polynomials

S. K. Crönert, *A Complete Methodology for the Predictive Simulation of Novel, Single- and Multi-Component Fuel Combustion*, Wissenschaftliche Reihe Fahrzeugtechnik Universität Stuttgart, https://doi.org/10.1007/978-3-658-43075-7

Table A.1: NASA polynomial coefficients of the used neat components for the low temperature regime (300 - 1000 K).

Fuel	Polynomial Coefficients								
	a_1	a_2	a_3	a_4	a_5	a_6	a_7	b_1	b_2
Isooctane	+1.93e0	−4.55e + 3	+2.77e + 1	+3.39e − 2	−1.70e − 6	+4.26e − 10	−5.28e − 14	−1.05e + 4	−1.32e + 2
Toluene	+1.75e0	−1.55e + 3	+6.98e0	+4.02e − 2	−1.19e − 5	−2.28e − 9	−5.85e − 14	+1.11e + 4	−1.77e + 1
n-Heptane	+2.58e − 1	−1.51e + 3	+9.43e0	+5.70e − 2	−1.73e − 5	−1.32e − 9	−1.01e − 13	−1.92e + 4	−2.36e + 1
Ethanol	−2.27e0	−1.07e + 3	+7.05e0	+1.61e − 2	−6.59e − 6	+1.93e − 9	−6.55e − 14	−2.49e + 4	−1.45e + 1
CPN	+1.38e0	−3.45e + 3	+1.73e + 1	+1.39e − 2	+1.96e − 7	+1.19e − 9	−5.58e − 14	−9.47e + 3	−7.70e + 1
AN	+2.78e0	−4.37e + 3	+2.47e + 1	+1.61e − 2	−2.20e − 6	+7.12e − 10	−4.19e − 14	+8.28e + 3	−1.18e + 2
DMC	+3.15e0	−3.19e + 2	+6.43e0	+2.84e − 2	−8.19e − 6	−9.18e − 10	−6.34e − 14	−6.96e + 4	−6.40e0
MeFo	+1.26e0	−4.87e + 2	+5.05e0	+1.71e − 2	−7.31e − 6	+1.93e − 9	−4.31e − 14	−4.13e + 4	−5.94e − 1
Hydrogen	+4.08e + 4	−8.01e + 2	+8.21e0	−1.27e − 2	+1.75e − 5	−1.20e − 8	+3.37e − 12	+2.68e + 3	−3.04e + 1

Table A.2: NASA polynomial coefficients of the used neat components for the high temperature regime (1000 - 3000 K).

Fuel	Polynomial Coefficients								
	a_1	a_2	a_3	a_4	a_5	a_6	a_7	b_1	b_2
Isooctane	+1.23e + 1	−1.39e + 4	+3.71e + 1	+5.36e − 2	−2.48e − 5	+2.16e − 9	+7.31e − 13	+4.21e + 4	−2.15e + 2
Toluene	+7.46e − 1	−1.65e + 4	+4.92e + 1	−5.53e − 4	−1.27e − 6	+6.67e − 10	−9.92e − 14	+8.80e + 4	−2.90e + 2
n-Heptane	+2.22e0	−3.18e + 4	+8.20e + 1	−1.13e − 3	−4.38e − 6	+1.36e − 9	+2.89e − 14	+1.42e + 5	−5.04e + 2
Ethanol	+2.30e0	−2.46e + 3	+1.24e + 1	+1.02e − 2	−3.62e − 6	+6.44e − 10	−5.40e − 14	−1.84e + 4	−4.82e + 1
CPN	+1.54e0	−9.57e + 3	+3.24e + 1	+8.44e − 3	−3.36e − 6	+6.99e − 10	−6.58e − 14	+2.18e + 4	−1.80e + 2
AN	+1.30e0	−9.03e + 3	+3.67e + 1	+1.02e − 2	−4.12e − 6	+7.79e − 10	−5.99e − 14	+3.20e + 4	−1.98e + 2
DMC	+3.01e − 2	+1.29e + 2	+1.30e + 1	+1.87e − 2	−7.25e − 6	+1.15e − 9	−4.68e − 14	−7.54e + 4	−4.29e + 1
MeFo	+1.22e0	+1.08e + 3	+4.59e0	+1.57e − 2	−6.05e − 6	+1.01e − 9	−5.55e − 14	−5.11e + 4	+5.28e0
Hydrogen	+5.61e + 5	−8.37e + 2	+2.98e0	+1.25e − 3	−3.74e − 7	+5.94e − 11	−3.61e − 15	+5.34e + 3	−2.20e0

Table A.3: Fuel mixtures NASA polynomial coefficients for the low temperature regime (300 - 1000 K).

Fuel	Polynomial Coefficients								
	a_1	a_2	a_3	a_4	a_5	a_6	a_7	b_1	b_2
TRF	+2.83e0	−8.32e + 2	+4.12e0	+6.66e − 2	−2.38e − 5	+1.02e − 10	+1.44e − 14	−1.47e + 4	+2.86e0
E5	+1.21e0	−3.49e + 3	+2.05e + 1	+3.06e − 2	−4.22e − 6	+1.57e − 10	−4.64e − 14	−4.50e + 3	−9.11e + 1
E10	−1.94e0	−6.39e + 2	+3.38e0	+5.79e − 2	−2.05e − 5	−6.22e − 11	−1.91e − 14	−1.80e + 4	+7.67e0
CPN50m	+3.24e0	−3.58e + 3	+1.95e + 1	+1.83e − 2	−1.94e − 7	+5.03e − 10	−3.83e − 14	−7.18e + 3	−8.70e + 1
AN50m	+2.76e0	−4.38e + 3	+2.53e + 1	+1.66e − 2	−2.49e − 7	+8.13e − 10	−4.14e − 14	+2.95e + 3	−1.19e + 2
CPN50v	+6.73e0	−8.77e + 2	+3.59e0	+4.43e − 2	−1.06e − 5	−3.15e − 9	−2.82e − 14	−1.92e + 4	+5.22e0
AN50v	+1.89e0	−4.24e + 3	+2.43e + 1	+1.86e − 2	−2.49e − 6	+1.40e − 9	−2.75e − 14	+3.09e + 3	−1.14e + 2
E5CPN	−1.37e0	−6.42e + 2	+2.76e0	+6.14e − 2	−2.29e − 5	+5.52e − 10	+4.11e − 14	−1.74e + 4	+1.07e + 1
E5AN	+1.12e − 1	−1.15e + 3	+6.67e0	+5.35e − 2	−1.42e − 5	−2.84e − 9	−9.65e − 14	−1.36e + 4	−1.07e + 1
E10CPN	+2.41e0	−4.09e + 3	+2.38e + 1	+1.92e − 2	−1.30e − 7	+1.00e − 9	−4.04e − 14	−3.31e + 3	−1.10e + 2
E10AN	−2.68e − 1	−6.42e + 2	+3.14e0	+5.85e − 2	−2.23e − 5	+4.90e − 10	+1.57e − 13	−1.71e + 4	+8.89e0
DMC65MeFo35	+1.26e0	−7.71e + 2	+8.13e0	+1.94e − 2	−5.46e − 6	+3.11e − 10	−6.46e − 14	−5.59e + 4	−1.64e + 1

Table A.4: Fuel mixtures NASA polynomial coefficients for the high temperature regime (1000 - 3000 K).

Fuel	Polynomial Coefficients								
	a_1	a_2	a_3	a_4	a_5	a_6	a_7	b_1	b_2
TRF	+1.97e0	−3.92e + 4	+9.55e + 1	−6.42e − 3	−6.07e − 6	+2.15e − 9	+4.16e − 14	+1.89e + 5	−6.03e + 2
E5	+3.88e − 1	−9.25e + 3	+2.62e + 1	+4.42e − 2	−2.18e − 5	+3.38e − 9	+1.68e − 13	+2.78e + 4	−1.42e + 2
E10	+3.17e0	−3.50e + 4	+8.61e + 1	−9.45e − 3	−3.59e − 6	+1.75e − 9	−1.61e − 14	+1.64e + 5	−5.40e + 2
CPN50m	+1.89e0	−1.80e + 4	+4.93e + 1	+7.11e − 3	−5.90e − 6	+1.44e − 9	−2.60e − 14	+7.00e + 4	−2.93e + 2
AN50m	+9.75e − 1	−1.62e + 4	+4.71e + 1	+1.33e − 2	−8.72e − 6	+1.89e − 9	−2.44e − 14	+6.71e + 4	−2.75e + 2
CPN50v	+2.18e0	−1.72e + 4	+4.79e + 1	+6.56e − 3	−5.31e − 6	+1.31e − 9	−3.13e − 14	+6.55e + 4	−2.84e + 2
AN50v	+1.33e0	−2.09e + 4	+5.89e + 1	+1.98e − 3	−4.44e − 6	+1.39e − 9	−5.68e − 14	+9.24e + 4	−3.52e + 2
E5CPN	+1.44e0	−1.45e + 4	+4.20e + 1	+2.31e − 2	−1.05e − 5	+6.70e − 10	+3.84e − 13	+5.39e + 4	−2.42e + 2
E5AN	+8.65e − 1	−3.43e + 4	+8.56e − 1	−6.97e − 3	−4.58e − 6	+1.93e − 9	−3.07e − 14	+1.62e + 5	−5.35e + 2
E10CPN	−1.11e0	−2.81e + 4	+6.98e + 1	+2.54e − 3	−7.71e − 6	+2.33e − 9	−4.62e − 14	+1.27e + 5	−4.32e + 2
E10AN	−1.60e − 1	−3.15e + 4	+7.92e + 1	−6.14e − 3	−3.88e − 6	+1.58e − 9	+1.96e − 15	+1.46e + 5	−4.92e + 2
DMC65MeFo35	+1.30e0	+1.08e + 3	+8.29e0	+1.81e − 2	−6.72e − 6	+9.77e − 10	−2.80e − 14	−6.80e + 4	−1.40e + 1

Appendix B: Mixture Definitions

Table B.1: Mixture definitions based on mass fractions. Volume based definitions in table 5.7, mole-fraction based definitions in table B.2.

Fuel(-mixture)	iso-Octane	toluene	n-heptane	ethanol	CPN	AN
	m.-%	m.-%	m.-%	m.-%	m.-%	m.-%
"Pure" Fuels						
T(P)RF	62.68	28.99	8.33	0	0	0
T(P)RF E5	59.32	27.44	7.88	5.36	0	0
T(P)RF E10	55.98	25.89	7.44	10.7	0	0
Cyclopentanone	0	0	0	0	100	0
Anisole	0	0	0	0	0	100
50 %						
$E10CPN_{50m}$	27.97	12.94	3.72	5.34	50.03	0
$E10AN_{50m}$	27.99	12.95	3.72	5.34	0	50
$E10CPN_{50V}$	24.49	11.33	3.25	4.67	56.25	0
$E10AN_{50V}$	23.85	11.03	3.17	4.55	0	57.39
EN228						
E5 CPN	53.73	24.69	7.09	4.82	10.03	0
E5 AN	51.39	23.77	6.83	4.64	0	13.37
5 wt.-% Oxygen						
E10 CPN	51.67	23.90	6.87	9.86	7.690	0
E10 AN	49.99	23.12	6.64	9.54	0	10.71

Table B.2: Mixture definitions based on molecular fractions at STP. Volume based definitions in table 5.7, mass fraction based definitions in table B.1.

Fuel(-mixture)	iso-Octane	toluen	n-heptane	ethanol	CPN	AN
	mol.-%	mol.-%	mol.-%	mol.-%	mol.-%	mol.-%
"Pure" Fuels						
T(P)RF	57.97	33.25	8.78	0.000	0	0
T(P)RF E5	51.31	29.42	7.77	11.50	0	0
T(P)RF E10	45.49	26.09	6.89	21.53	0	0
Cyclopentanone	0	0	0	0	100	0
Anisole	0	0	0	0	0	100
50 %						
$E10CPN_{50m}$	21.61	12.39	3.27	10.23	52.49	0
$E10AN_{50m}$	24.48	14.04	3.71	11.59	0	46.19
$E10CPN_{50V}$	18.81	10.78	2.85	8.900	58.66	0
$E10AN_{50V}$	21.10	12.10	3.20	9.990	0	53.62
EN228						
E5 CPN	45.37	26.02	6.87	10.17	11.58	0
E5 AN	44.97	25.79	6.81	10.08	0	12.35
5 wt.-% Oxygen						
E10 CPN	41.66	23.89	6.31	19.72	8.42	0
E10 AN	41.24	23.65	6.25	19.52	0	9.33

Appendix C: Transport Properties

Table C.1: Dynamic viscosities of the pure fuels.

Fuel	Temperature [K]								
	250	280	310	340	370	400	430	460	490
Isooctane	+5.01e − 6	+5.62e − 6	+6.24e − 6	+6.86e − 6	+7.49e − 6	+8.12e − 6	+8.74e − 6	+9.36e − 6	+9.97e − 6
Toluene	+5.53e − 6	+6.19e − 6	+6.86e − 6	+7.54e − 6	+8.23e − 6	+8.92e − 6	+9.61e − 6	+1.03e − 5	+1.10e − 5
n-Heptane	+4.93e − 6	+5.53e − 6	+6.14e − 6	+6.75e − 6	+7.37e − 6	+7.99e − 6	+8.60e − 6	+9.21e − 6	+9.81e − 6
Ethanol	+6.64e − 6	+7.45e − 6	+8.26e − 6	+9.09e − 6	+9.92e − 6	+1.08e − 5	+1.16e − 5	+1.24e − 5	+1.32e − 5
CPN	+6.06e − 6	+6.79e − 6	+7.53e − 6	+8.28e − 6	+9.03e − 6	+9.79e − 6	+1.05e − 5	+1.13e − 5	+1.20e − 5
AN	+5.61e − 6	+6.24e − 6	+6.90e − 6	+7.57e − 6	+8.26e − 6	+8.95e − 6	+9.64e − 6	+1.03e − 5	+1.10e − 5
DMC	+6.55e − 6	+7.29e − 6	+8.06e − 6	+8.84e − 6	+9.64e − 6	+1.04e − 5	+1.13e − 5	+1.21e − 5	+1.29e − 5
MeFo	+9.83e − 6	+1.11e − 5	+1.24e − 5	+1.36e − 5	+1.48e − 5	+1.61e − 5	+1.73e − 5	+1.85e − 5	+1.96e − 5
Hydrogen	+7.88e − 6	+8.51e − 6	+9.12e − 6	+9.70e − 6	+1.03e − 5	+1.08e − 5	+1.13e − 5	+1.18e − 5	+1.23e − 5

Table C.2: Thermal conductivities of the mixtures.

Fuel	Temperature [K]								
	250	280	310	340	370	400	430	460	490
TRF	+5.16e − 6	+5.78e − 6	+6.42e − 6	+7.06e − 6	+7.70e − 6	+8.35e − 6	+8.99e − 6	+9.63e − 6	+1.03e − 5
TRFE5	+5.27e − 6	+5.91e − 6	+6.56e − 6	+7.22e − 6	+7.88e − 6	+8.54e − 6	+9.19e − 6	+9.85e − 6	+1.05e − 5
TRFE10	+5.38e − 6	+6.03e − 6	+6.69e − 6	+7.37e − 6	+8.04e − 6	+8.71e − 6	+9.38e − 6	+1.00e − 5	+1.07e − 5
E10CPN50m	+5.72e − 6	+6.41e − 6	+7.11e − 6	+7.82e − 6	+8.53e − 6	+9.25e − 6	+9.96e − 6	+1.07e − 5	+1.14e − 5
E10AN50m	+5.50e − 6	+6.14e − 6	+6.80e − 6	+7.47e − 6	+8.15e − 6	+8.83e − 6	+9.51e − 6	+1.02e − 5	+1.09e − 5
E10CPN50v	+5.76e − 6	+6.45e − 6	+7.16e − 6	+7.87e − 6	+8.59e − 6	+9.32e − 6	+1.00e − 5	+1.07e − 5	+1.15e − 5
E10AN50v	+5.51e − 6	+6.15e − 6	+6.81e − 6	+7.49e − 6	+8.17e − 6	+8.85e − 6	+9.53e − 6	+1.02e − 5	+1.09e − 5
E5CPN	+5.35e − 6	+6.00e − 6	+6.66e − 6	+7.32e − 6	+7.99e − 6	+8.66e − 6	+9.33e − 6	+9.99e − 6	+1.06e − 5
E5AN	+5.32e − 6	+5.95e − 6	+6.60e − 6	+7.26e − 6	+7.93e − 6	+8.59e − 6	+9.25e − 6	+9.91e − 6	+1.06e − 5
E10CPN	+5.43e − 6	+6.09e − 6	+6.76e − 6	+7.43e − 6	+8.11e − 6	+8.79e − 6	+9.47e − 6	+1.01e − 5	+1.08e − 5
E10AN	+5.41e − 6	+6.06e − 6	+6.72e − 6	+7.39e − 6	+8.06e − 6	+8.74e − 6	+9.41e − 6	+1.01e − 5	+1.07e − 5

Table C.3: Thermal conductivities of the pure fuels.

Fuel	Temperature [K]								
	250	280	310	340	370	400	430	460	490
Isooctane	+1.31e − 2	+1.30e − 2	+1.42e − 2	+1.63e − 2	+1.91e − 2	+2.23e − 2	+2.58e − 2	+2.96e − 2	+3.35e − 2
Toluene	+7.41e − 3	+8.97e − 3	+1.09e − 2	+1.32e − 2	+1.57e − 2	+1.83e − 2	+2.11e − 2	+2.40e − 2	+2.70e − 2
n-Heptane	+1.09e − 2	+1.21e − 2	+1.40e − 2	+1.64e − 2	+1.92e − 2	+2.23e − 2	+2.56e − 2	+2.90e − 2	+3.26e − 2
Ethanol	+1.17e − 2	+1.37e − 2	+1.62e − 2	+1.90e − 2	+2.21e − 2	+2.54e − 2	+2.89e − 2	+3.25e − 2	+3.63e − 2
CPN	+7.30e − 3	+8.61e − 3	+1.04e − 2	+1.26e − 2	+1.51e − 2	+1.79e − 2	+2.08e − 2	+2.39e − 2	+2.71e − 2
AN	+7.50e − 3	+9.03e − 3	+1.09e − 2	+1.31e − 2	+1.54e − 2	+1.80e − 2	+2.06e − 2	+2.34e − 2	+2.62e − 2
DMC	+1.05e − 2	+1.16e − 2	+1.32e − 2	+1.52e − 2	+1.74e − 2	+1.98e − 2	+2.24e − 2	+2.51e − 2	+2.79e − 2
MeFo	+1.42e − 2	+1.60e − 2	+1.84e − 2	+2.12e − 2	+2.44e − 2	+2.78e − 2	+3.13e − 2	+3.51e − 2	+3.89e − 2
Hydrogen	+1.56e − 1	+1.68e − 1	+1.81e − 1	+1.93e − 1	+2.05e − 1	+2.16e − 1	+2.28e − 1	+2.40e − 1	+2.51e − 1

Table C.4: Dynamic viscosities of the mixtures.

Fuel	Temperature [K]								
	250	280	310	340	370	400	430	460	490
TRF	+1.07e − 2	+1.14e − 2	+1.30e − 2	+1.52e − 2	+1.79e − 2	+2.09e − 2	+2.41e − 2	+2.75e − 2	+3.11e − 2
TRFE5	+1.08e − 2	+1.17e − 2	+1.33e − 2	+1.56e − 2	+1.83e − 2	+2.14e − 2	+2.46e − 2	+2.81e − 2	+3.16e − 2
TRFE10	+1.09e − 2	+1.19e − 2	+1.36e − 2	+1.60e − 2	+1.87e − 2	+2.18e − 2	+2.51e − 2	+2.86e − 2	+3.21e − 2
E10CPN50m	+8.85e − 3	+1.00e − 2	+1.18e − 2	+1.41e − 2	+1.67e − 2	+1.96e − 2	+2.28e − 2	+2.60e − 2	+2.94e − 2
E10AN50m	+9.18e − 3	+1.05e − 2	+1.23e − 2	+1.46e − 2	+1.71e − 2	+1.99e − 2	+2.29e − 2	+2.60e − 2	+2.92e − 2
E10CPN50v	+8.64e − 3	+9.85e − 3	+1.17e − 2	+1.39e − 2	+1.65e − 2	+1.94e − 2	+2.25e − 2	+2.57e − 2	+2.91e − 2
E10AN50v	+8.94e − 3	+1.03e − 2	+1.21e − 2	+1.43e − 2	+1.69e − 2	+1.97e − 2	+2.26e − 2	+2.56e − 2	+2.88e − 2
E5CPN	+1.03e − 2	+1.13e − 2	+1.30e − 2	+1.52e − 2	+1.79e − 2	+2.09e − 2	+2.42e − 2	+2.76e − 2	+3.11e − 2
E5AN	+1.03e − 2	+1.13e − 2	+1.30e − 2	+1.53e − 2	+1.79e − 2	+2.09e − 2	+2.41e − 2	+2.74e − 2	+3.09e − 2
E10CPN	+1.05e − 2	+1.16e − 2	+1.33e − 2	+1.56e − 2	+1.84e − 2	+2.14e − 2	+2.47e − 2	+2.81e − 2	+3.17e − 2
E10AN	+1.05e − 2	+1.16e − 2	+1.34e − 2	+1.57e − 2	+1.84e − 2	+2.14e − 2	+2.46e − 2	+2.80e − 2	+3.15e − 2

Appendix D: LBV Correlation Fits

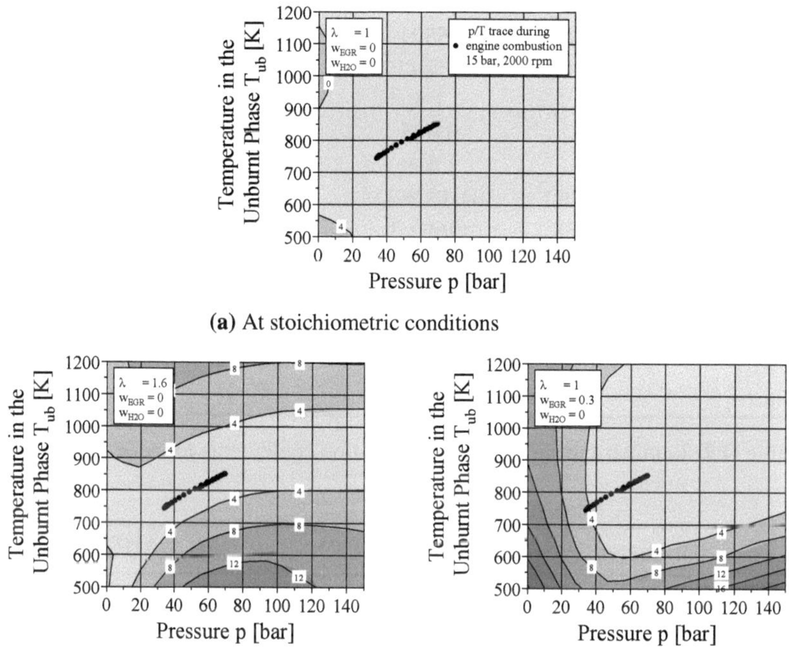

(a) At stoichiometric conditions

(b) At lean conditions (λ = 1.6)

(c) With extended EGR ratio

Figure D.1: Laminar burning velocity fit quality for methyl formate with varying boundary conditions.

Table D.1: Laminar burning velocity correlation coefficients for added neat fuels.

parameter	Anisole	Cyclopentanone	Methanol	Methyl Formate
E_i	46984	55521	48996	50933
B_i	6.15e+17	6.15e+17	1.23e+18	1.23e+18
m	1.5	1.5	1.5	1.5
r	0.985	0.985	0.985	0.985
n	2.439	2.439	2.439	2.439
F	0.0259392	0.0122822	0.0094399	0.001266
G	-6369.28	-9692.3	-7781.5	-10501.9
n_{EGR}	1.8063666	1.2473852	1.41346	1.21657
n_a	0.94549	0.94549	1.01992	1.01992
$s_{3\text{-fac}}$	0.74789	0.74789	0.72148	0.82011
b_{H2O}	0.2106823	-0.015846	0.60224	0.892592
c_{H2O}	0	0	0	0

Table D.2: Laminar burning velocity correlation coefficients for added fuel mixtures.

parameter	$E10AN_{50V}$	$E10CPN_{50V}$	$DMC_{65}MF_{35V}$
E_i	54998	54541	56158
B_i	6.15 e+17	6.1488 e+17	6.1488 e+17
m	1.5	1.5	1.5
r	0.985	0.985	0.985
n	2.439	2.439	2.439
F	0.0544038	0.0078898	0.01812
G	-7169.22	-9792.39	-8610.31
n_{EGR}	0.9407556	0.9435724	1.03794
n_a	0.94549	0.94549	0.94549
$s_{3\text{-fac}}$	0.74789	0.74789	0.74789
b_{H2O}	0.2106827*	0.2106827*	0.97566
c_{H2O}	0	0	0

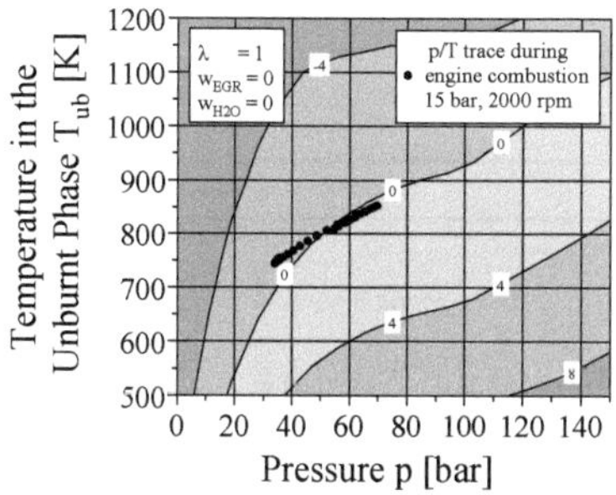

(a) At stoichiometric conditions

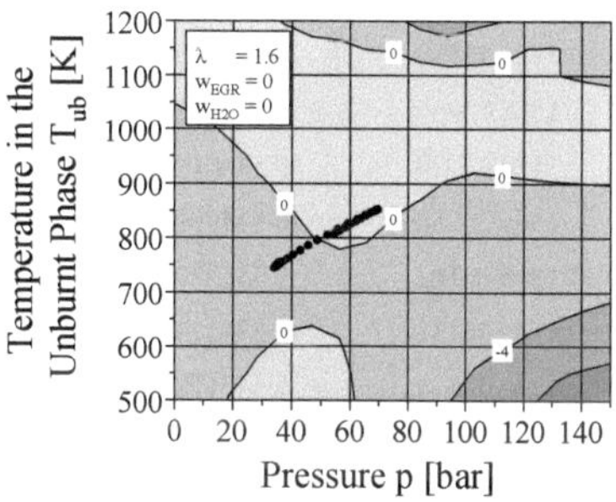

(b) At lean conditions ($\lambda = 1.6$)

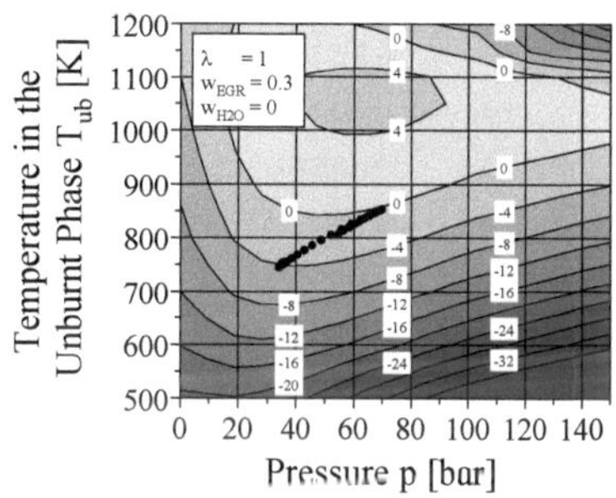

(c) With extended EGR ratio

Figure D.2: Laminar burning velocity fit quality for DMC with varying boundary conditions.

Table D.3: Spline data for the laminar burning velocity calculation of Anisole.

λ	S_1	S_2	S_3	S_4
0.1	-0.11519196	0.82253003	643.952663	0.52158914
0.2	-0.10497588	0.87676044	750.669509	0.55002736
0.3	-0.09367574	0.94377314	897.063866	0.59658751
0.4	-0.08303054	0.99856408	1065.74003	0.65337497
0.5	-0.07185431	1.04952062	1227.81055	0.70127822
0.6	-0.06079899	1.04447121	1304.42135	0.73034475

Continued on next page

Table D.3: Spline data for the laminar burning velocity calculation of Anisole.(Continued)

λ	S_1	S_2	S_3	S_4
0.7	-0.04897605	1.02708707	1452.29813	0.78631810
0.8	-0.03837281	0.99017427	1693.10692	0.64632910
0.9	-0.03855702	1.00291744	1876.24423	0.58104737
1.0	-0.01595013	0.99330157	2000.62263	0.49411972
1.1	-0.00481018	0.98888276	1932.57220	0.49403474
1.2	0.00637594	0.98872492	1833.73639	0.57835253
1.3	0.01753029	0.96682382	1725.73552	0.53845587
1.4	0.02880870	0.96866449	1568.83466	0.53831963
1.5	0.04086412	0.94571074	1520.07020	0.54101986
1.6	0.05423953	0.93775394	1505.30050	0.54490120
1.7	0.06846435	0.92370496	1365.63163	0.53133015
1.8	0.08144849	0.92233404	1307.76151	0.53841296
1.9	0.09434567	0.92774747	1230.24814	0.54719889
2.0	0.10749330	0.91457294	1168.17928	0.55203213
2.1	0.12034762	0.90576978	1124.07500	0.56125889
2.2	0.13328390	0.89459361	1076.82868	0.56583844
2.3	0.14878597	0.88316338	1039.02543	0.57386129
2.4	0.15784477	0.82340709	1017.97658	0.58742850
2.5	0.17184002	0.86378048	970.799867	0.59105012
2.6	0.18490835	0.85391193	937.439606	0.60632263
2.7	0.19806859	0.83549311	921.719331	0.61369325
2.8	0.20828197	0.79038935	897.552200	0.57108026
2.9	0.22289697	0.78283657	867.549123	0.63000226
3.0	0.23556417	0.76833242	843.822566	0.63785308
3.1	0.24863929	0.75992322	828.289200	0.65179040
3.2	0.26135728	0.74072578	794.575971	0.65513179
3.3	0.27448699	0.72602937	770.025674	0.66286908
3.4	0.28791135	0.71184097	747.884380	0.67072135
3.5	0.29977040	0.69663805	719.382767	0.68221013
3.6	0.31063349	0.67936218	686.156515	0.69385365

Continued on next page

Table D.3: Spline data for the laminar burning velocity calculation of Anisole.(Continued)

λ	S_1	S_2	S_3	S_4
3.7	0.32293330	0.66429435	660.974711	0.70432292
3.8	0.33697583	0.66172070	670.675491	0.70587095
3.9	0.34971825	0.64937309	653.777664	0.71411991
4.0	0.36207944	0.63496512	631.427359	0.72384311
4.1	0.37732008	0.63413738	644.124016	0.72752144
4.2	0.39067319	0.62358399	632.695891	0.73424388
4.3	0.40358172	0.61086204	615.335478	0.74253766
4.4	0.41512812	0.59338275	584.481359	0.75465642
4.5	0.42945303	0.58560826	580.750842	0.76017204
4.6	0.44664530	0.58599585	603.834918	0.75567714
4.7	0.46035029	0.57423422	590.603738	0.76256683
4.8	0.47257049	0.55855521	565.980723	0.77259396
4.9	0.49002351	0.55288554	573.183377	0.77366399
5.0	0.50699991	0.55164567	585.723591	0.77786751

Table D.4: Spline data for the laminar burning velocity calculation of Cyclopentanone.

λ	S_1	S_2	S_3	S_4
0.1	-0.11519196	0.82253003	643.952663	0.52158914
0.2	-0.10497588	0.87676044	750.669509	0.55002736
0.3	-0.09367574	0.94377314	897.063866	0.59658751
0.4	-0.08303054	0.99856408	1065.74003	0.65337497
0.5	-0.07180379	1.02309591	1236.73860	0.71038561
0.6	-0.05702818	1.00429244	1358.14600	0.75196204
0.7	-0.04971146	1.00016970	1488.33224	0.79986951

Continued on next page

Table D.4: Spline data for the laminar burning velocity calculation of Cyclopentanone.(Continued)

λ	S_1	S_2	S_3	S_4
0.8	-0.03853917	0.99026952	1725.81183	0.65035487
0.9	-0.03840856	0.99873802	1897.12651	0.58066917
1.0	-0.01599978	0.98937617	1947.89605	0.50755742
1.1	-0.00483033	0.98553410	1917.73354	0.49301289
1.2	0.00621462	0.98180703	1796.53035	0.57915386
1.3	0.01756892	0.96973949	1710.85784	0.54055968
1.4	0.02865973	0.95848844	1602.38537	0.55448963
1.5	0.04076340	0.94639216	1522.90521	0.54644838
1.6	0.05361968	0.93119481	1448.44376	0.52920408
1.7	0.06839556	0.91921110	1385.66493	0.53796988
1.8	0.08087880	0.90724732	1325.02353	0.54642339
1.9	0.09318865	0.89580016	1273.89992	0.55393293
2.0	0.10539262	0.88321619	1221.88986	0.56362434
2.1	0.12267260	0.86925656	1177.98404	0.57538731
2.2	0.13074991	0.85679690	1131.10979	0.58262970
2.3	0.14448882	0.83923531	1084.44920	0.58857200
2.4	0.15672235	0.82440144	1046.77651	0.59506066
2.5	0.16940283	0.80774270	1010.63811	0.59837332
2.6	0.18401177	0.81979381	954.373542	0.60490620
2.7	0.19688313	0.80699068	923.994106	0.61299207
2.8	0.20969527	0.79406834	895.089336	0.62145187
2.9	0.22258843	0.78120768	868.716320	0.62985577
3.0	0.23556417	0.76833242	843.822566	0.63785308
3.1	0.24863929	0.75992322	828.289200	0.65179040
3.2	0.26135728	0.74072578	794.575971	0.65513179
3.3	0.27448699	0.72602937	770.025674	0.66286908
3.4	0.28791135	0.71184097	747.884380	0.67072135
3.5	0.29977040	0.69663805	719.382760	0.68221013
3.6	0.31063349	0.67936218	686.156515	0.69385365
3.7	0.32293330	0.66429435	660.974711	0.70432292

Continued on next page

Table D.4: Spline data for the laminar burning velocity calculation of Cyclopentanone.(Continued)

λ	S_1	S_2	S_3	S_4
3.8	0.33697583	0.6617207	670.675491	0.70587095
3.9	0.34971825	0.64937309	653.777664	0.71411991
4.0	0.36207944	0.63496512	631.427359	0.72384311
4.1	0.37732008	0.63413738	644.124016	0.72752144
4.2	0.39067319	0.62358399	632.695891	0.73424388
4.3	0.40358172	0.61086204	615.335478	0.74253766
4.4	0.41512812	0.59338275	584.481359	0.75465642
4.5	0.42945303	0.58560826	580.750842	0.76017204
4.6	0.44664530	0.58599585	603.834918	0.75567714
4.7	0.46035029	0.57423422	590.603738	0.76256683
4.8	0.47257049	0.55855521	565.980723	0.77259396
4.9	0.49002351	0.55288554	573.183377	0.77366399
5.0	0.50699991	0.55164567	585.723591	0.77786751

Table D.5: Spline data for the laminar burning velocity calculation of the $E10AN_{50V}$ mixture.

λ	S_1	S_2	S_3	S_4
0.1	-0.11519196	0.82253003	643.952663	0.52158914
0.2	-0.10497588	0.87676044	750.669509	0.55002736
0.3	-0.09367574	0.94377314	897.063866	0.59658751
0.4	-0.08303054	0.99856408	1065.74003	0.65337497
0.5	-0.07180379	1.02309591	1236.73860	0.71038561
0.6	-0.06069047	1.01189727	1335.62353	0.74687206
0.7	-0.04953057	1.00532073	1470.63396	0.79378784
0.8	-0.03737341	0.99890024	1698.05256	0.64395394

Continued on next page

Table D.5: Spline data for the laminar burning velocity calculation of the $E10AN_{50V}$ mixture.(Continued)

λ	S_1	S_2	S_3	S_4
0.9	-0.03854077	1.00427393	1880.28464	0.58555789
1.0	-0.01606645	0.99367681	1962.70250	0.50935333
1.1	-0.00483161	0.98631478	1923.01132	0.49284217
1.2	0.00617617	0.98691108	1838.82000	0.58709363
1.3	0.01751986	0.96977217	1713.63181	0.53705285
1.4	0.02855981	0.95678486	1615.49358	0.55537846
1.5	0.04057120	0.94324281	1534.84559	0.54610752
1.6	0.05355644	0.92963027	1448.51762	0.55544135
1.7	0.06806921	0.91404477	1388.24093	0.53964127
1.8	0.08074393	0.90348538	1328.04445	0.54720198
1.9	0.09291744	0.89064093	1277.09822	0.55056910
2.0	0.10571712	0.87628989	1224.44301	0.56404316
2.1	0.11985151	0.88084443	1146.33711	0.56497150
2.2	0.13267497	0.86929678	1101.83893	0.57287324
2.3	0.14550502	0.85734576	1060.77248	0.58080627
2.4	0.15833918	0.84506026	1022.71272	0.58877972
2.5	0.17119103	0.83286061	987.286231	0.59672301
2.6	0.18401177	0.81979381	954.373542	0.60490620
2.7	0.19688313	0.80699068	923.994106	0.61299207
2.8	0.20969527	0.79406834	895.089336	0.62145187
2.9	0.22258843	0.78120768	868.716320	0.62985577
3.0	0.23556417	0.76833242	843.822566	0.63785308
3.1	0.24863929	0.75992322	828.289200	0.65179040
3.2	0.26135728	0.74072578	794.575971	0.65513179
3.3	0.27448699	0.72602937	770.025674	0.66286908
3.4	0.28791135	0.71184097	747.884380	0.67072135
3.5	0.29977040	0.69663805	719.382767	0.68221013
3.6	0.31063349	0.67936218	686.156515	0.69385365
3.7	0.32293330	0.66429435	660.974711	0.70432292
3.8	0.33697583	0.66172070	670.675491	0.70587095

Continued on next page

Table D.5: Spline data for the laminar burning velocity calculation of the $E10AN_{50V}$ mixture.(Continued)

λ	S_1	S_2	S_3	S_4
3.9	0.34971825	0.64937309	653.777664	0.71411991
4.0	0.36207944	0.63496512	631.427359	0.72384311
4.1	0.37732008	0.63413738	644.124016	0.72752144
4.2	0.39067319	0.62358399	632.695891	0.73424388
4.3	0.40358172	0.61086204	615.335478	0.74253766
4.4	0.41512812	0.59338275	584.481359	0.75465642
4.5	0.42945303	0.58560826	580.750842	0.76017204
4.6	0.44664530	0.58599585	603.834918	0.75567714
4.7	0.46035029	0.57423422	590.603738	0.76256683
4.8	0.47257049	0.55855521	565.980723	0.77259396
4.9	0.49002351	0.55288554	573.183377	0.77366399
5.0	0.50699991	0.55164567	585.723591	0.77786751

Table D.6: Spline data for the laminar burning velocity calculation of the $E10CPN_{50V}$ mixture.

λ	S_1	S_2	S_3	S_4
0.1	-0.11519196	0.82253003	643.952662	0.52158914
0.2	-0.10497588	0.87676044	750.669509	0.55002736
0.3	-0.09367574	0.94377314	897.063865	0.59658751
0.4	-0.08303054	0.99856408	1065.74002	0.65337497
0.5	-0.07180379	1.02309591	1236.73859	0.71038561
0.6	-0.06069047	1.01189727	1335.62352	0.74687206
0.7	-0.04953057	1.00532073	1470.63396	0.79378784
0.8	-0.03822263	1.00097250	1689.32503	0.64323245
0.9	-0.03839792	1.00354405	1876.02595	0.58109321

Continued on next page

Table D.6: Spline data for the laminar burning velocity calculation of the E10CPN$_{50V}$ mixture.(Continued)

λ	S_1	S_2	S_3	S_4
1.0	-0.01605161	0.99218048	1950.48495	0.50757633
1.1	-0.0048144	0.98647245	1925.80286	0.49377425
1.2	0.00633856	0.98435051	1839.13445	0.58017735
1.3	0.01751986	0.96977217	1713.63180	0.53705285
1.4	0.02856995	0.95724657	1609.20823	0.55347713
1.5	0.04074183	0.94533608	1524.92256	0.54724864
1.6	0.05398761	0.93226325	1445.17067	0.55092695
1.7	0.06798665	0.91819455	1382.90278	0.53838657
1.8	0.08117474	0.90751295	1323.05304	0.54645700
1.9	0.09321646	0.89422900	1268.04875	0.55444885
2.0	0.10577989	0.88179455	1220.23482	0.56374114
2.1	0.11985151	0.88084443	1146.33711	0.56497150
2.2	0.13267497	0.86929678	1101.83893	0.57287324
2.3	0.14550502	0.85734576	1060.77247	0.58080627
2.4	0.15833918	0.84506026	1022.71271	0.58877972
2.5	0.17119103	0.83286061	987.286231	0.59672301
2.6	0.18401177	0.81979381	954.373541	0.60490620
2.7	0.19688313	0.80699068	923.994106	0.61299207
2.8	0.20969527	0.79406834	895.089335	0.62145187
2.9	0.22258843	0.78120768	868.716320	0.62985577
3.0	0.23556417	0.76833242	843.822566	0.63785308
3.1	0.24863929	0.75992322	828.289200	0.65179040
3.2	0.26135728	0.74072578	794.575970	0.65513179
3.3	0.27448699	0.72602937	770.025673	0.66286908
3.4	0.28791135	0.71184097	747.884380	0.67072135
3.5	0.29977040	0.69663805	719.382766	0.68221013
3.6	0.31063349	0.67936218	686.156515	0.69385365
3.7	0.32293330	0.66429435	660.974710	0.70432292
3.8	0.33697583	0.66172070	670.675491	0.70587095
3.9	0.34971825	0.64937309	653.777664	0.71411991

Continued on next page

Table D.6: Spline data for the laminar burning velocity calculation of the $E10CPN_{50V}$ mixture.(Continued)

λ	S_1	S_2	S_3	S_4
4.0	0.36207944	0.63496512	631.427359	0.72384311
4.1	0.37732008	0.63413738	644.124015	0.72752144
4.2	0.39067319	0.62358399	632.695890	0.73424388
4.3	0.40358172	0.61086204	615.335477	0.74253766
4.4	0.41512812	0.59338275	584.481358	0.75465642
4.5	0.42945303	0.58560826	580.750842	0.76017204
4.6	0.44664530	0.58599585	603.834918	0.75567714
4.7	0.46035029	0.57423422	590.603737	0.76256683
4.8	0.47257049	0.55855521	565.980722	0.77259396
4.9	0.49002351	0.55288554	573.183377	0.77366399
5.0	0.50699991	0.55164567	585.723590	0.77786751

Appendix E: Hydrogen Combustion

Due to the special characteristics of hydrogen combustion, the laminar burning velocity correlation approach presented in subsection 2.3.3 cannot be applied. A new method based on a polynomial equation E.1 for the pressure dependence of the burning velocity is created. All other influences are then only applied by multiplying factors onto the base value (equations E.2 to E.5) ultimately leading to equation E.6. This way, all multipliers equal one if the boundary conditions are at the reference point, what simplifies debugging and maintenance. All corresponding values for the used calibration variables are given in E.1. The splines are still used for the burning velocity dependence on the fuel equivalence ratio but reduced to three and repurposed in their application. Their values are shown in table E.2.

$$s_{\mathrm{L,0}} = p_1 p_{\mathrm{cyl}}^4 + p_2 p_{\mathrm{cyl}}^3 + p_3 p_{\mathrm{cyl}}^2 + p_4 p_{\mathrm{cyl}} + p_5 \tag{E.1}$$

$$mult_T = T_{\mathrm{a}} e^{(T_{\mathrm{ex}} \cdot T_{\mathrm{ub}})} \tag{E.2}$$

$$mult_{\mathrm{lam}} = S_1 \cdot (S_2 \cdot p_{\mathrm{cyl}}^{S_3}) \tag{E.3}$$

$$mult_{\mathrm{EGR}} = e^{(p_{\mathrm{egr,log}} \cdot \log p + p_{\mathrm{egr,add}}) \cdot -w_{\mathrm{EGR}}} \tag{E.4}$$

$$mult_{\mathrm{H2O}} = 1 + h2o_{\mathrm{a}}(w_{\mathrm{EGR}} + h2o_{\mathrm{b}}) w_{\mathrm{H2O}} \tag{E.5}$$

$$s_{\mathrm{L}} = s_{\mathrm{L,0}} \cdot mult_T \cdot mult_{\mathrm{lam}} \cdot mult_{\mathrm{EGR}} \cdot mult_{\mathrm{H2O}} \tag{E.6}$$

Table E.1: Laminar burning velocity correlation coefficients for hydrogen.

Parameter	Value
p_1	2.3944 e-8
p_2	-1.3130 e-5
p_3	2.7425 e-3
p_4	-0.27998
p_5	14.3686
T_a	2.71 e-2
T_{ex}	4.5 e-3
$p_{egr,log}$	1.1939
$p_{egr,add}$	0.6423
$h2o_a$	-0.1672
$h2o_b$	-7.9 e-3

Table E.2: Spline data for the laminar burning velocity calculation of Hydrogen.

λ	S_1	S_2	S_3
0.1	0.1135477872811	4.4968559308027	-0.3410761144416
0.2	0.6916407040864	1.1315189324521	-0.0263927020299
0.3	1.1357718947758	0.7380856806398	0.08564472058572
0.4	1.3830269654786	0.6297128607252	0.12607164196873
0.5	1.4883800326855	0.6014341157328	0.13791256394308
0.6	1.4887000934004	0.6118856192176	0.13320468410597
0.7	1.4292261325002	0.6426911571596	0.11885989971916
0.8	1.3140622522731	0.7091334624195	0.09241424837130
0.9	1.1632207722154	0.8284177803364	0.05068283142248
1.0	1.0000000000000	1.0073583659000	-0.0002813253860
1.1	0.8495994465714	1.2020437808814	-0.0466875256867
1.2	0.7137077590527	1.4671352560044	-0.0960663249506
1.3	0.6013606815270	1.7258337318107	-0.1382107669838

Continued on next page

Table E.2: Spline data for the laminar burning velocity calculation of Hydrogen.(Continued)

λ	S_1	S_2	S_3
1.4	0.5007756399977	2.0734392578349	-0.1825052338685
1.5	0.4213019872691	2.4576735179663	-0.2261568707451
1.6	0.3451513333056	2.9673991616058	-0.2689288943995
1.7	0.2892946164294	3.4991921354779	-0.3101978958331
1.8	0.2393159757480	4.1763348700045	-0.3527175640967
1.9	0.1975240013260	5.0215261063041	-0.3973284158622
2.0	0.1634421814794	6.0246967936336	-0.4424509417854
2.1	0.1369134326001	6.7734662083234	-0.4744109070538
2.2	0.1134685440521	7.8222003626728	-0.5080967602189
2.3	0.0922381927547	9.2000876620876	-0.5427071757840
2.4	0.0770218838229	10.451482031511	-0.5740219570226
2.5	0.0638275189046	11.848525898779	-0.6026636006469
2.6	0.0525888935198	13.336531762446	-0.6270728383057
2.7	0.0432476861810	14.573403308207	-0.6425565373662
2.8	0.0361386419275	15.706817165491	-0.6579304436530
2.9	0.0306432341424	16.426564130754	-0.6675496793098
3.0	0.0262504480010	16.414282806000	-0.6678330100209
3.1	0.0226323676111	16.013052179574	-0.6625155368952
3.2	0.0194667906977	15.325446831918	-0.6493164036932
3.3	0.0169941063600	14.542344543084	-0.6354408235870
3.4	0.0148365896835	12.829708379017	-0.6032087348240
3.5	0.0132588987583	10.891855105513	-0.5647788529359
3.6	0.0120033767052	8.4161363296191	-0.5033714820335
3.7	0.0106844963557	7.8129354663541	-0.4830012873140
3.8	0.0096583947256	6.3892862612441	-0.4342013044781
3.9	0.0086995103495	5.2912939829929	-0.3863422452208
4.0	0.0080351909136	4.3022986126950	-0.3374289773601
4.1	0.0072253261645	3.7351786963708	-0.2994661488471
4.2	0.0065777632954	3.3590749966574	-0.2708932377191
4.3	0.0060607008602	3.1817118505985	-0.2578764775712

Continued on next page

Table E.2: Spline data for the laminar burning velocity calculation of Hydrogen.(Continued)

λ	S_1	S_2	S_3
4.4	0.0055814459933	2.9237982741514	-0.2369678674222
4.5	0.0052099447361	2.6926552158023	-0.2187011377088
4.6	0.0048189798672	2.4361385334667	-0.1935212832132
4.7	0.0045499487260	2.3058698801169	-0.1833258439093
4.8	0.0042012600202	2.1421968759201	-0.1627142748411
4.9	0.0039061260926	1.9427663171808	-0.1365144945816
5.0	0.0036344603266	1.8172501653070	-0.1174709358986
5.1	0.0034122577422	1.7481831001680	-0.1064579928524
5.2	0.0032110135554	1.6264266982190	-0.0876600236058
5.3	0.0030286212383	1.5698598186689	-0.0790894748975
5.4	0.0028722465336	1.4877520527866	-0.0653498396342
5.5	0.0027239149442	1.4175048461086	-0.0517107743667
5.6	0.0025813094558	1.3336013448766	-0.0354982870073
5.7	0.0024501650301	1.3072920961632	-0.0299628252245
5.8	0.0023373820063	1.2581666476829	-0.0208698197502
5.9	0.0022247508589	1.2543935612359	-0.0196614401867
6.0	0.0020500000000	1.2506204747890	-0.0184530606231
6.1	0.0019325377001	1.2468473883421	-0.0172446810596
6.2	0.0018221807712	1.2430743018951	-0.0160363014961
6.3	0.0017181257385	1.2393012154482	-0.0148279219325
6.4	0.0016200127340	1.2355281290012	-0.0136195423690
6.5	0.0015275024404	1.2317550425543	-0.0124111628054
6.6	0.0014402749167	1.2279819561074	-0.0112027832419
6.7	0.0013580284919	1.2242088696604	-0.0099944036783
6.8	0.0012804787222	1.2204357832135	-0.0087860241148
6.9	0.0012073574065	1.2166626967665	-0.0075776445513
7.0	0.0011384116595	1.2128896103196	-0.0063692649877
7.1	0.0010734030366	1.2091165238727	-0.0051608854242
7.2	0.0010121067097	1.2053434374257	-0.0039525058606
7.3	0.0009543106894	1.2015703509788	-0.0027441262971

Continued on next page

Table E.2: Spline data for the laminar burning velocity calculation of Hydrogen.(Continued)

λ	S_1	S_2	S_3
7.4	0.0008998150918	1.1977972645318	-0.0015357467335
7.5	0.0008484314474	1.1940241780849	-0.0003273671700
7.6	0.0007999820490	1.1902510916380	0.00088101239349
7.7	0.0007542993376	1.1864780051910	0.00208939195703
7.8	0.0007112253224	1.1827049187441	0.00329777152057
7.9	0.0006706110346	1.1789318322971	0.00450615108412
8.0	0.0006323160123	1.1751587458502	0.00571453064766
8.1	0.0005962078147	1.1713856594033	0.00692291021120
8.2	0.0005621615638	1.1676125729563	0.00813128977474
8.3	0.0005300595129	1.1638394865094	0.00933966933829
8.4	0.0004997906390	1.1600664000624	0.01054804890183
8.5	0.0004712502591	1.1562933136155	0.01175642846537
8.6	0.0004443396683	1.1525202271686	0.01296480802891
8.7	0.0004189657979	1.1487471407216	0.01417318759246
8.8	0.0003950408940	1.1449740542747	0.01538156715600
8.9	0.0003724822138	1.1412009678277	0.01658994671954
9.0	0.0003512117397	1.1374278813808	0.01779832628308
9.1	0.0003311559090	1.1336547949339	0.01900670584663
9.2	0.0003122453599	1.1298817084869	0.02021508541017
9.3	0.0002944146914	1.1261086220400	0.02142346497371
9.4	0.0002776022374	1.1223355355930	0.02263184453725
9.5	0.0002617498531	1.1185624491461	0.02384022410080
9.6	0.0002468027140	1.1147893626992	0.02504860366434
9.7	0.0002327091263	1.1110162762522	0.02625698322788
9.8	0.0002194203482	1.1072431898053	0.02746536279142
9.9	0.0002068904215	1.1034701033584	0.02867374235497
10	0.00019507601219	1.0996970169114	0.02988212191851

Appendix F: Ignition Delay Time Fits

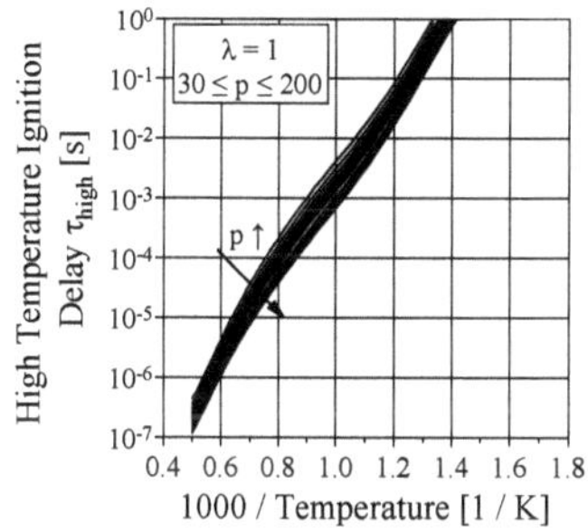

(a) Toluene

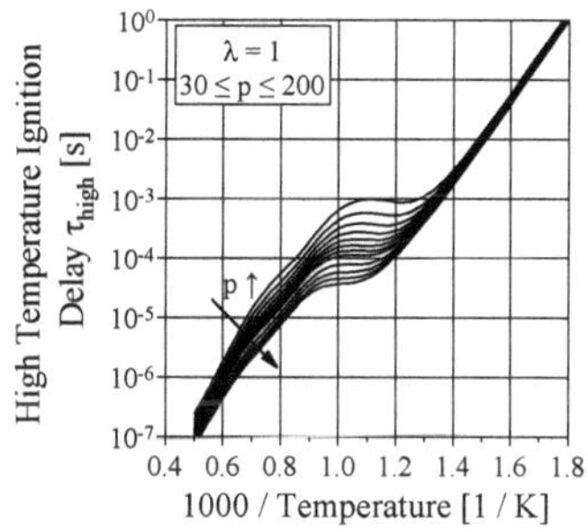

(b) n-Heptane

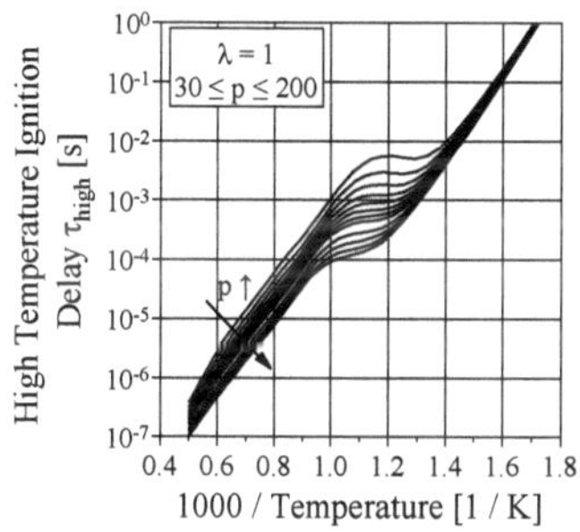

(c) iso-Octane

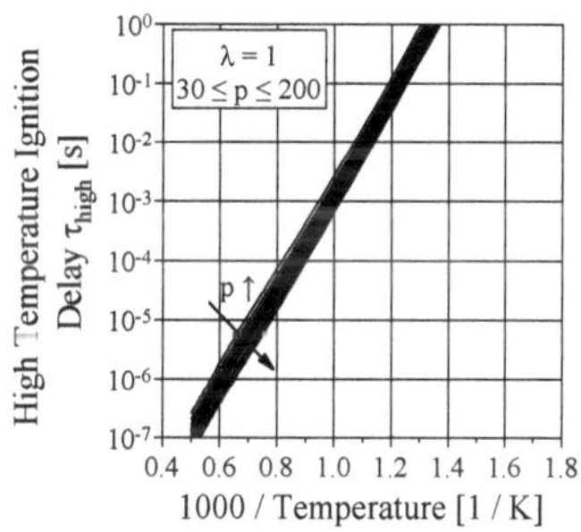

(d) Methanol

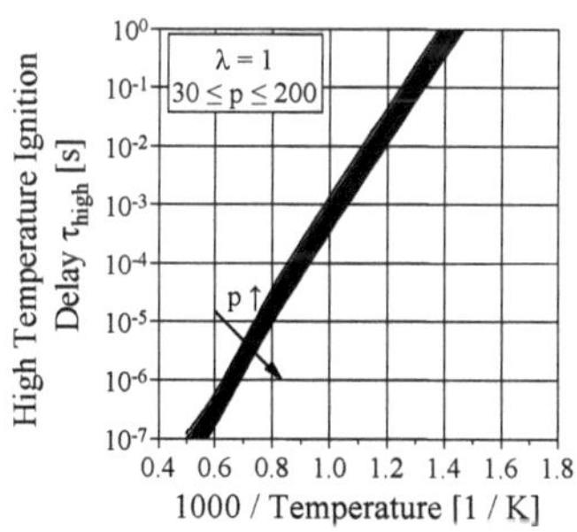

(e) Ethanol

Figure F.1: Comparison of the high temperature ignition delays of multiple pure fuel components at stoichiometric conditions and varying pressures.

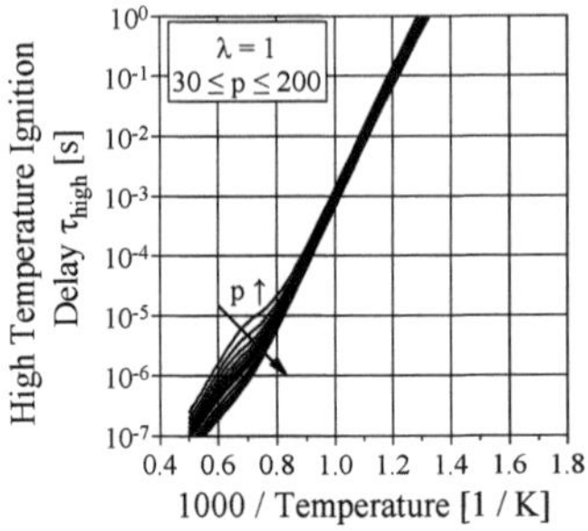

(f) Anisole

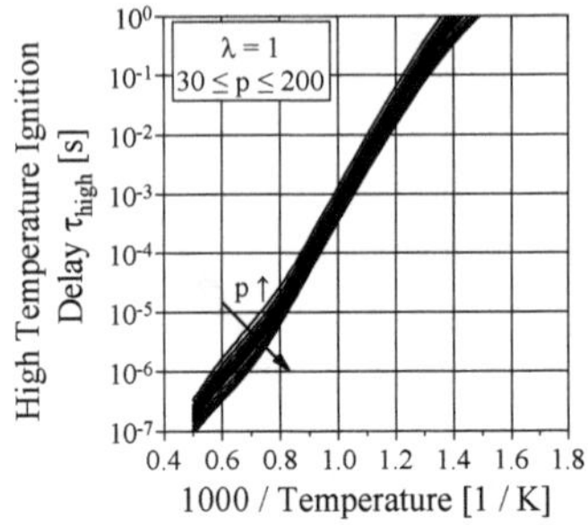

(g) Cyclopentanone

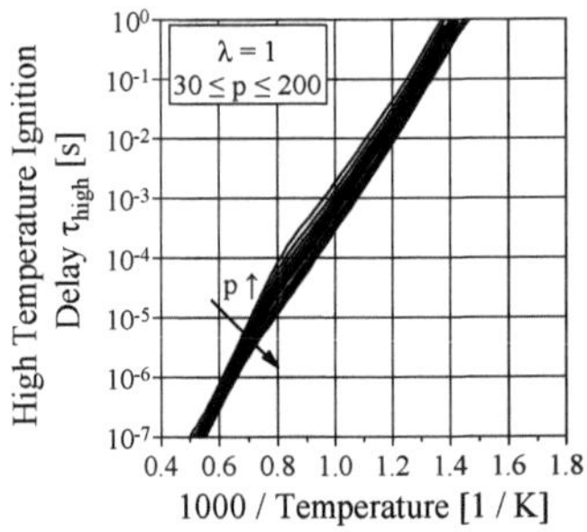

(h) Dimethyl Carbonate

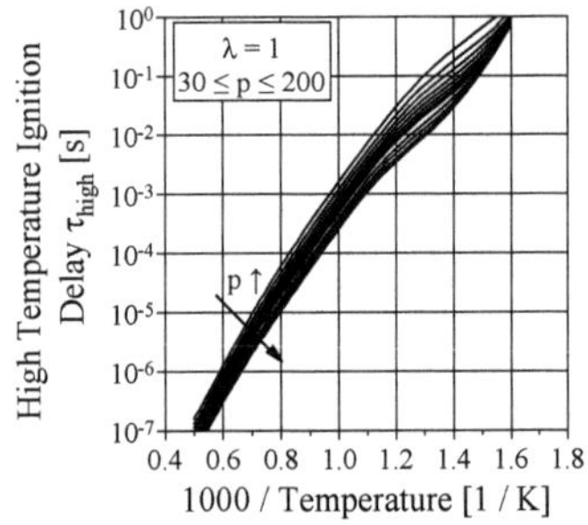

(i) Methyl Formate

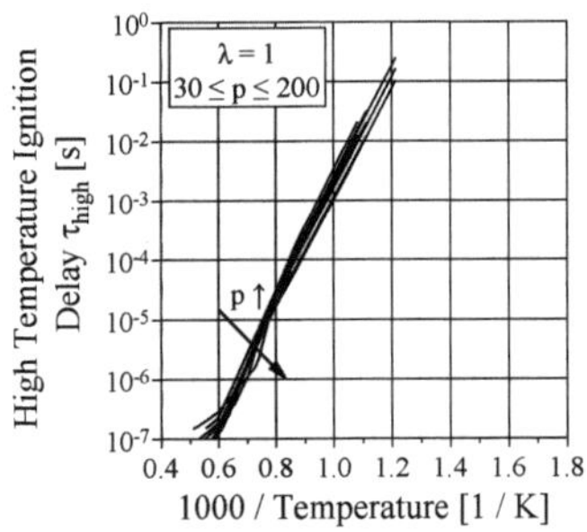

(j) Hydrogen

Figure F.1: Comparison of the high temperature ignition delays of multiple pure fuel components at stoichiometric conditions and varying pressures (continued).

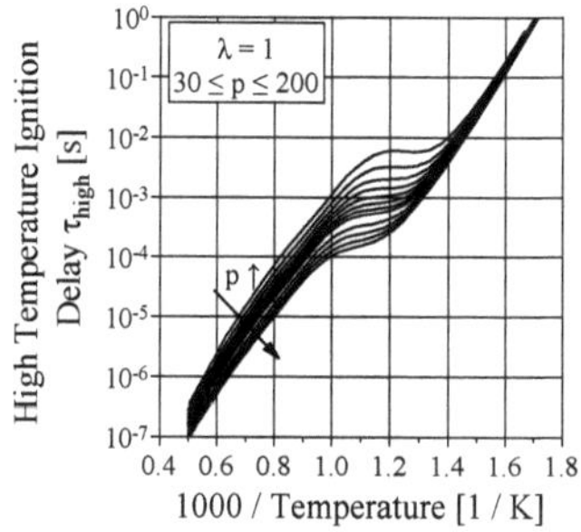

(a) TRFE5

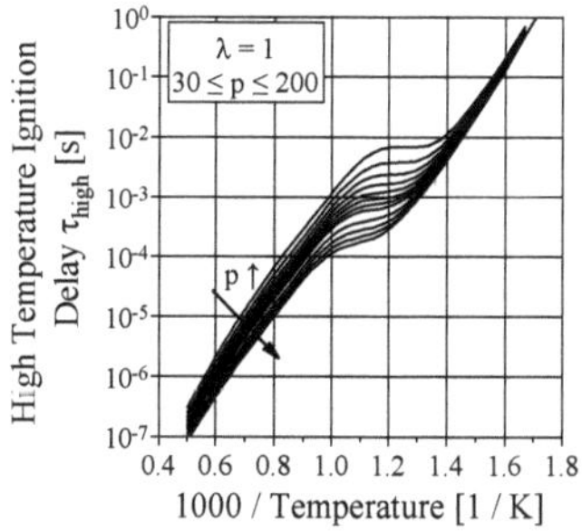

(b) TRFE10

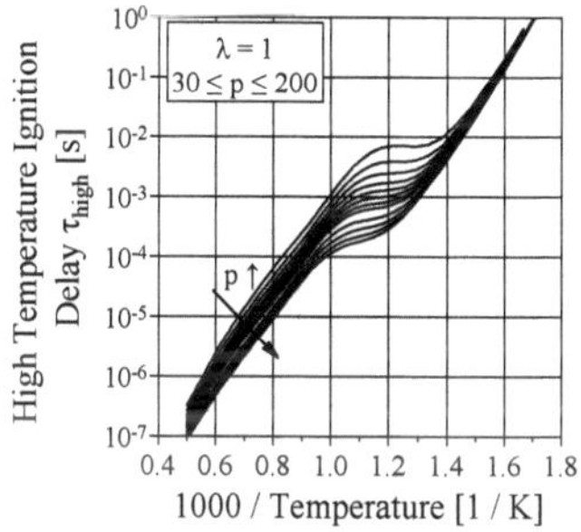

(c) E5 AN

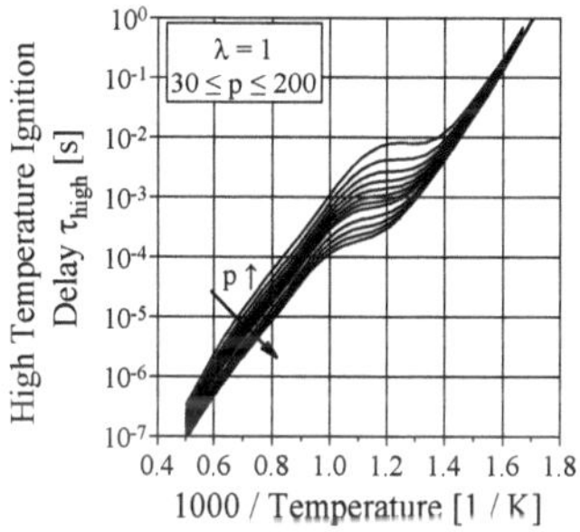

(d) E5 CPN

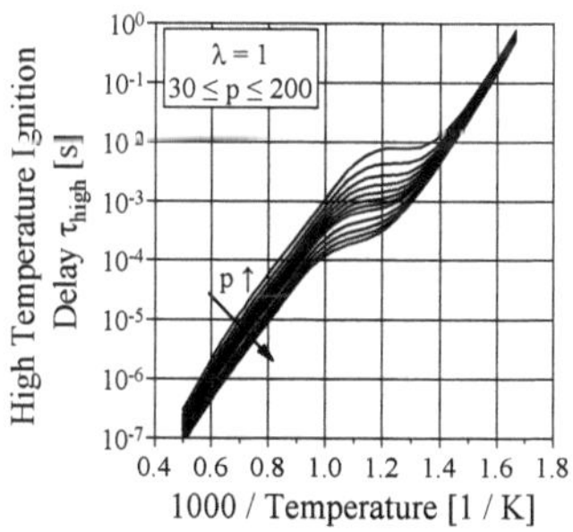

(e) E10 AN

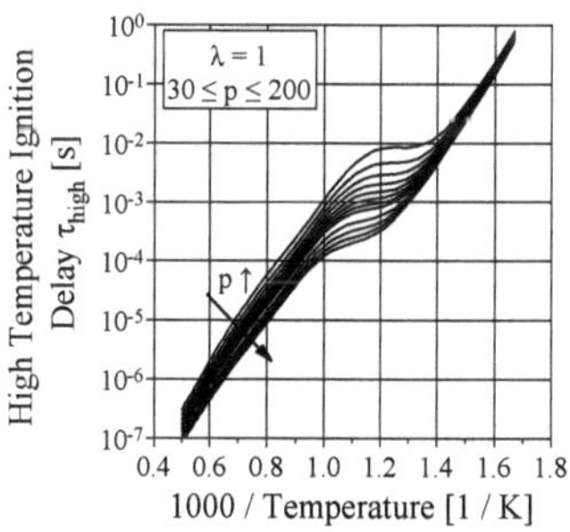

(f) E10 CPN

Figure F.2: Comparison of the high temperature ignition delays of multiple fuel mixtures at stoichiometric conditions and varying pressures.

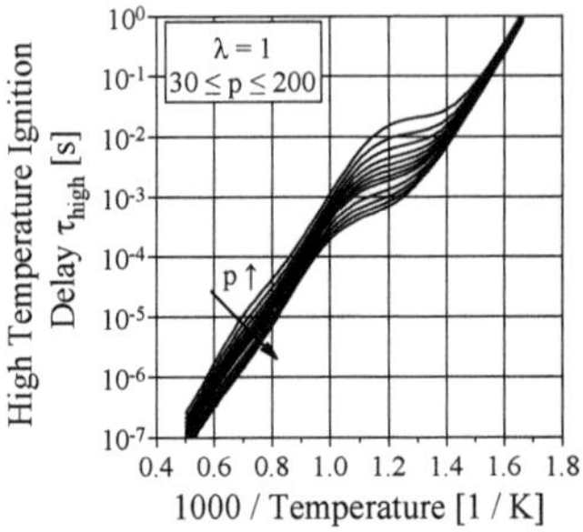

(g) $E10AN_{50m}$

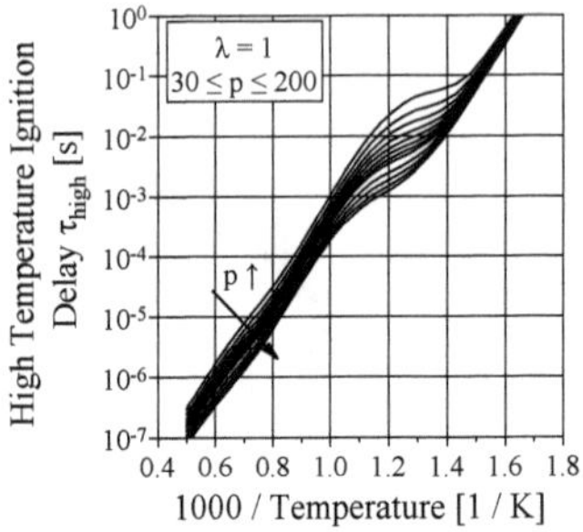

(h) $E10CPN_{50m}$

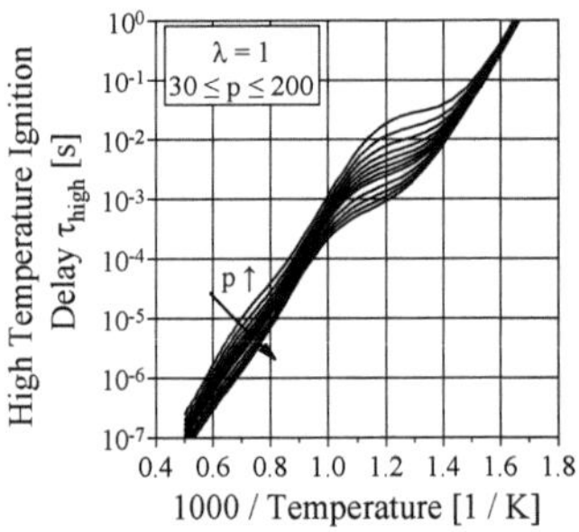

(i) $E10AN_{50V}$

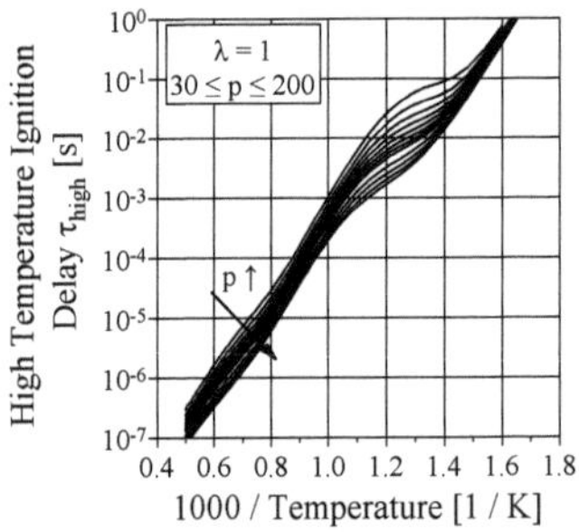

(j) $E10CPN_{50V}$

Figure F.2: Comparison of the high temperature ignition delays of multiple fuel mixtures at stoichiometric conditions and varying pressures (continued).

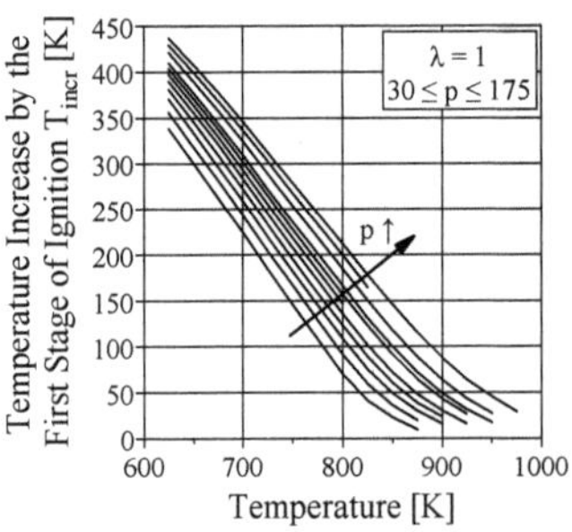

(a) iso-Octane

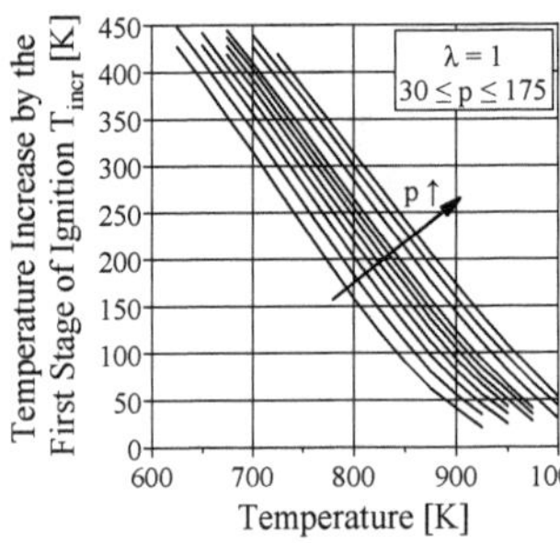

(b) n-Heptane

Figure F.3: Temperature increase due to the first stage of ignition for iso-Octane and n-Heptane at stoichiometric conditions and varying pressures.

Table F.1: High temperature ignition delay correlation coefficients for the $E10AN_{50V}$ mixture.

Parameter	Coefficients				
			$\tau_{high,1}$		
$U_{A,1}$	1.93E-07	-7.31E-05	9.90E-03	-6.01E-01	-1.39E+01
$P_{A,1}$	6.94E-06	-1.67E-03	9.49E-02	7.53E-01	
$Q_{A,1}$	-1.39E-05	4.22E-03	-3.90E-01	1.30E+01	
$R_{A,1}$	6.62E-09	-2.23E-06	2.20E-04	-7.35E-03	
$S_{A,1}$	-4.90E-07	1.72E-04	-1.93E-02	8.63E-01	
$U_{B,1}$	-1.27E-04	4.79E-02	-6.43E+00	3.83E+02	8.00E+03
$P_{B,1}$	-4.83E-03	1.18E+00	-7.13E+01	-9.67E+01	
$Q_{B,1}$	9.17E-03	-2.77E+00	2.54E+02	-8.18E+03	
$R_{B,1}$	-4.32E-06	1.46E-03	-1.44E-01	4.68E+00	
$S_{B,1}$	3.09E-04	-1.08E-01	1.21E+01	-5.26E+02	
			$\tau_{high,2}$		
$U_{A,2}$	1.80E-07	-6.93E-05	9.82E-03	-6.83E-01	1.85E+01
$P_{A,2}$	6.35E-06	-1.48E-03	7.12E-02	1.50E+00	
$Q_{A,2}$	-1.10E-05	3.09E-03	-2.29E-01	-3.63E-01	
$R_{A,2}$	4.34E-09	-1.44E-06	1.33E 04	-4.16E-03	
$S_{A,2}$	-4.08E-07	1.41E-04	-1.53E-02	6.30E-01	
$U_{B,2}$	-1.42E-04	5.31E-02	-7.17E+00	4.55E+02	-1.53E+04
$P_{B,2}$	-5.65E-03	1.33E+00	-6.77E+01	-1.31E+03	
$Q_{B,2}$	9.63E-03	-2.74E+00	2.10E+02	6.13E+02	
$R_{B,2}$	-3.67E-06	1.22E-03	-1.10E-01	2.57E+00	
$S_{B,2}$	3.26E-04	-1.13E-01	1.20E+01	-4.07E+02	
			$\tau_{high,3}$		
$U_{A,3}$	1.38E-08	-5.93E-06	1.01E-03	-9.52E-02	-2.05E+01
$P_{A,3}$	4.13E-07	-1.22E-04	9.91E-03	2.92E-01	
$Q_{A,3}$	-3.86E-07	8.95E-05	4.18E-03	-9.69E-01	
$R_{A,3}$	2.03E-10	-7.16E-08	6.60E-06	-1.26E-03	
$S_{A,3}$	-4.54E-09	1.51E-06	-2.66E-05	1.00E-01	
$U_{B,3}$	-1.14E-05	4.38E-03	-6.38E-01	5.14E+01	1.47E+04
$P_{B,3}$	-5.31E-04	1.57E-01	-1.30E+01	-3.57E+02	
$Q_{B,3}$	6.41E-04	-1.77E-01	5.39E+00	1.37E+03	
$R_{B,3}$	-2.52E-07	8.56E-05	-6.64E-03	8.99E-01	
$S_{B,3}$	5.92E-06	-1.75E-03	-5.49E-02	-6.11E+01	

Table F.2: High temperature ignition delay correlation coefficients for the $E10CPN_{50V}$ mixture.

Parameter	Coefficients				
			$\tau_{high,1}$		
$U_{A,1}$	1.71E-20	-5.90E-18	6.88E-16	-3.12E-14	-1.72E+03
$P_{A,1}$	2.03E-06	-3.91E-04	1.40E-02	1.21E+00	
$Q_{A,1}$	-3.30E-06	1.11E-03	-1.39E-01	2.12E+03	
$R_{A,1}$	1.33E-05	2.35E-03	-1.18E+00	4.68E+01	
$S_{A,1}$	-1.99E-04	-3.54E-02	1.77E+01	-7.02E+02	
$U_{B,1}$	-1.25E-03	4.79E-01	-6.50E+01	3.66E+03	1.00E+06
$P_{B,1}$	7.61E-04	-4.39E-01	7.62E+01	-3.68E+05	
$Q_{B,1}$	-8.82E-09	2.59E-06	-2.34E-04	-1.00E+06	
$R_{B,1}$	-8.34E-03	-1.45E+00	7.35E+02	-2.92E+04	
$S_{B,1}$	1.25E-01	2.19E+01	-1.10E+04	4.38E+05	
			$\tau_{high,2}$		
$U_{A,2}$	1.27E-22	-5.38E-20	7.64E-18	-4.16E-16	-1.27E+01
$P_{A,2}$	-2.66E-06	7.33E-04	-6.61E-02	2.85E+00	
$Q_{A,2}$	-2.68E-07	3.88E-04	-1.14E-01	1.95E+01	
$R_{A,2}$	1.76E-08	3.45E-05	-8.52E-03	3.12E-01	
$S_{A,2}$	5.78E-08	-7.01E-04	1.57E-01	-5.62E+00	
$U_{B,2}$	1.84E-19	-6.49E-17	7.80E-15	-3.69E-13	7.32E+03
$P_{B,2}$	1.90E-03	-5.18E-01	4.63E+01	-2.38E+03	
$Q_{B,2}$	-9.89E-04	1.93E-01	8.22E+00	-1.17E+04	
$R_{B,2}$	-7.07E-06	-3.41E-02	8.33E+00	-3.01E+02	
$S_{B,2}$	-7.30E-05	6.11E-01	-1.41E+02	5.03E+03	
			$\tau_{high,3}$		
$U_{A,3}$	-3.01E-22	1.22E-19	-1.78E-17	1.07E-15	-2.42E+01
$P_{A,3}$	-7.95E-08	1.76E-05	-1.16E-03	3.88E-01	
$Q_{A,3}$	-3.32E-07	1.49E-04	-2.60E-02	1.20E+00	
$R_{A,3}$	2.51E-08	-8.53E-06	1.09E-03	-3.11E-02	
$S_{A,3}$	-3.36E-07	9.17E-05	-9.30E-03	2.25E-01	
$U_{B,3}$	4.16E-20	-2.50E-17	4.60E-15	-3.08E-13	1.73E+04
$P_{B,3}$	9.10E-05	-1.77E-02	7.27E-01	-5.53E+02	
$Q_{B,3}$	-3.04E-04	1.07E-01	-1.64E+01	4.79E+02	
$R_{B,3}$	-2.05E-05	4.12E-03	-8.87E-02	-3.54E+00	
$S_{B,3}$	2.93E-04	-4.18E-02	-3.01E+00	2.05E+02	

Table F.3: Low temperature ignition delay correlation coefficients for the E10AN$_{50V}$ mixture.

Parameter	Coefficients			
		$\tau_{low,1}$		
$U_{A,1}$	-9.24E-06	3.22E-03	-4.07E-01	-1.23E+00
$P_{A,1}$	5.82E-05	-1.70E-02	1.48E+00	-3.90E+01
$Q_{A,1}$	-4.34E-05	1.26E-02	-1.15E+00	3.59E+01
$R_{A,1}$	NaN*	NaN*	NaN*	NaN*
$S_{A,1}$	NaN*	NaN*	NaN*	NaN*
$U_{B,1}$	6.50E-03	-2.25E+00	2.85E+02	-3.18E+03
$P_{B,1}$	-6.44E-02	1.87E+01	-1.63E+03	4.33E+04
$Q_{B,1}$	3.38E-02	-9.76E+00	8.91E+02	-2.75E+04
$R_{B,1}$	NaN*	NaN*	NaN*	NaN*
$S_{B,1}$	NaN*	NaN*	NaN*	NaN*
		$\tau_{low,2}$		
$U_{A,2}$	-3.21E-06	1.08E-03	-1.35E-01	-2.58E+01
$P_{A,2}$	5.04E-05	-1.36E-02	1.08E+00	-2.65E+01
$Q_{A,2}$	-2.56E-05	7.19E-03	-6.18E-01	1.79E+01
$R_{A,2}$	NaN*	NaN*	NaN*	NaN*
$S_{A,2}$	NaN*	NaN*	NaN*	NaN*
$U_{B,2}$	1.89E-03	-6.30E-01	7.77E+01	1.56E+04
$P_{B,2}$	-2.89E-02	7.73E+00	-6.02E+02	1.44E+04
$Q_{B,2}$	1.68E-02	-4.73E+00	4.08E+02	-1.16E+04
$R_{B,2}$	NaN*	NaN*	NaN*	NaN*
$S_{B,2}$	NaN*	NaN*	NaN*	NaN*

Table F.4: Low temperature ignition delay correlation coefficients for the E10CPN$_{50V}$ mixture.

Parameter	Coefficients			
	$\tau_{low,1}$			
$U_{A,1}$	9.91E-05	-1.81E-02	1.05E+00	-3.46E+01
$P_{A,1}$	-1.85E-03	3.01E-01	-1.59E+01	2.70E+02
$Q_{A,1}$	5.57E-03	-9.05E-01	4.78E+01	-8.16E+02
$R_{A,1}$	NaN*	NaN*	NaN*	NaN*
$S_{A,1}$	NaN*	NaN*	NaN*	NaN*
$U_{B,1}$	-7.63E-02	1.41E+01	-8.43E+02	2.35E+04
$P_{B,1}$	2.24E+00	-3.65E+02	1.93E+04	-3.33E+05
$Q_{B,1}$	-4.08E+00	6.62E+02	-3.49E+04	5.95E+05
$R_{B,1}$	NaN*	NaN*	NaN*	NaN*
$S_{B,1}$	NaN*	NaN*	NaN*	NaN*
	$\tau_{low,2}$			
$U_{A,2}$	2.13E-05	-3.97E-03	2.38E-01	-3.45E+01
$P_{A,2}$	1.84E-03	-3.02E-01	1.62E+01	-2.85E+02
$Q_{A,2}$	-5.72E-03	9.36E-01	-5.01E+01	8.74E+02
$R_{A,2}$	NaN*	NaN*	NaN*	NaN*
$S_{A,2}$	NaN*	NaN*	NaN*	NaN*
$U_{B,2}$	-1.27E-02	2.39E+00	-1.47E+02	2.12E+04
$P_{B,2}$	-2.24E+00	3.66E+02	-1.95E+04	3.39E+05
$Q_{B,2}$	4.19E+00	-6.85E+02	3.66E+04	-6.38E+05
$R_{B,2}$	NaN*	NaN*	NaN*	NaN*
$S_{B,2}$	NaN*	NaN*	NaN*	NaN*

Table F.5: Correlation coefficients for the E10AN$_{50V}$ mixture temperature increase by low-temperature pre-reactions.

Parameter	Coefficients				
			C_1		
$P_{C,1}$	9.70E-03	-2.01E+00	9.11E+01		
$Q_{C,1}$	6.66E-04	1.48E-01	8.05E+00		
$R_{C,1}$	-1.43E-07	5.94E-05	-5.78E-04		
$S_{C,1}$	5.21E-07	-1.25E-04	-8.85E-03		
$U_{C,1}$	-2.64E-09	-2.64E-06	1.02E-03	-1.24E-01	7.42E+00
			C_2		
$P_{C,2}$	-2.81E-01	5.82E+01	-2.63E+03		
$Q_{C,2}$	-1.94E-02	4.33E+00	-2.39E+02		
$R_{C,2}$	1.71E-06	-9.42E-04	5.78E-02		
$S_{C,2}$	1.14E-06	-5.09E-04	1.12E-02		
$U_{C,2}$	2.87E-07	-7.22E-06	-1.75E-02	2.80E+00	-1.90E+02
			C_3		
$P_{C,3}$	3.04E+00	-6.30E+02	2.84E+04		
$Q_{C,3}$	2.12E-01	-4.74E+01	2.65E+03		
$R_{C,3}$	-3.51E-05	1.24E-02	-5.84E-01		
$S_{C,3}$	1.02E-05	-9.69E-05	-3.87E-02		
$U_{C,3}$	-5.22E-06	9.14E-04	6.88E-02	-2.25E+01	1.80E+03
			C_4		
$P_{C,4}$	-1.46E+01	3.01E+03	-1.36E+05		
$Q_{C,4}$	-1.03E+00	2.30E+02	-1.29E+04		
$R_{C,4}$	2.17E-04	-6.38E-02	2.71E+00		
$S_{C,4}$	6.96E-05	-2.05E-02	6.10E-01		
$U_{C,4}$	3.43E-05	-8.04E-03	1.90E-01	7.52E+01	-7.64E+03
			C_5		
$P_{C,5}$	2.60E+01	-5.37E+03	2.42E+05		
$Q_{C,5}$	1.88E+00	-4.21E+02	2.33E+04		
$R_{C,5}$	-5.30E-04	1.33E-01	-5.61E+00		
$S_{C,5}$	-2.00E-04	4.61E-02	-1.46E+00		
$U_{C,5}$	-7.70E-05	2.05E-02	-1.20E+00	-8.20E+01	1.26E+04

Table F.6: Correlation coefficients for the E10CPN$_{50V}$ mixture temperature increase by low-temperature pre-reactions.

Parameter	Coefficients				
			C_1		
$P_{C,1}$	-8.30E-03	1.83E+00	-7.43E+01		
$Q_{C,1}$	1.87E-03	-3.30E-01	1.70E+01		
$R_{C,1}$	-7.86E-06	1.46E-03	-5.41E-02		
$S_{C,1}$	-3.50E-06	6.28E-04	-3.65E-02		
$U_{C,1}$	2.81E-07	-1.09E-04	1.54E-02	-9.39E-01	2.29E+01
			C_2		
$P_{C,2}$	2.26E-01	-4.83E+01	2.02E+03		
$Q_{C,2}$	-5.48E-02	9.64E+00	-4.95E+02		
$R_{C,2}$	2.33E-04	-4.33E-02	1.69E+00		
$S_{C,2}$	9.19E-05	-1.67E-02	6.00E-01		
$U_{C,2}$	-7.80E-06	3.02E-03	-4.23E-01	2.57E+01	-6.18E+02
			C_3		
$P_{C,3}$	-2.28E+00	5.07E+02	-2.05E+04		
$Q_{C,3}$	5.97E-01	-1.05E+02	5.35E+03		
$R_{C,3}$	-2.571E-03	4.78E-01	-1.86E+01		
$S_{C,3}$	-8.60E-04	1.60E-01	-6.31E+00		
$U_{C,3}$	8.11E-05	-3.13E-02	4.37E+00	-2.63E+02	6.24E+03
			C_4		
$P_{C,4}$	1.02E+01	-2.29E+03	9.25E+04		
$Q_{C,4}$	-2.87E+00	5.04E+02	-2.55E+04		
$R_{C,4}$	1.22E-02	-2.27E+00	8.82E+01		
$S_{C,4}$	5.13E-03	-9.74E-01	3.91E+01		
$U_{C,3}$	-3.74E-04	1.44E-01	-2.00E+01	1.20E+03	-2.82E+04
			C_5		
$P_{C,5}$	-1.70E+01	3.86E+03	-1.56E+05		
$Q_{C,5}$	5.16E+00	-9.02E+02	4.51E+04		
$R_{C,5}$	-2.21E-02	4.11E+00	-1.61E+02		
$S_{C,5}$	-8.94E-03	1.67E+00	-6.48E+01		
$U_{C,5}$	6.74E-04	-2.48E-01	3.44E+01	-2.06E+03	4.82E+04

MIX
Papier aus verantwortungsvollen Quellen
Paper from responsible sources
FSC® C105338

If you have any concerns about our products,
you can contact us on
ProductSafety@springernature.com

In case Publisher is established outside the EU,
the EU authorized representative is:
Springer Nature Customer Service Center GmbH
Europaplatz 3, 69115 Heidelberg, Germany

Printed by Libri Plureos GmbH
in Hamburg, Germany